Valentin Bryantsev

# The effect of solar activity in marine ecosystems

Valentin Bryantsev

# The effect of solar activity in marine ecosystems

ScienciaScripts

Cover image: www.ingimage.com

This book is a translation from the original published under ISBN 978-613-3-99256-6.

Publisher:
Sciencia Scripts
is a trademark of
Dodo Books Indian Ocean Ltd. and OmniScriptum S.R.L publishing group

120 High Road, East Finchley, London, N2 9ED, United Kingdom
Str. Armeneasca 28/1, office 1, Chisinau MD-2012, Republic of Moldova, Europe
Printed at: see last page
**ISBN: 978-620-8-07762-4**

# CONTENTS.

# INTRODUCTION

In the process of oceanological research (Southern Research Institute of Marine Fisheries and Oceanography, Kerch, Ukraine) to improve the methodology of long-term forecasts of catch of fish and seafood in different areas of the World Ocean, the links between the states of marine ecosystems and long-term changes in the characteristics of the atmosphere and hydrosphere, which conditioned the physical prerequisites of the aquatic environment, yield and peculiarities of the behavior of fishing objects, were revealed.

Such quasi-periodic changes include, for example, the auto-oscillatory processes in the North Atlantic and South Atlantic Oscillation systems, the El Niño phenomenon created by periodic intrusions of waters of the Equatorial Countercurrent into the Peru-Chile upwelling zone, and a number of others.

However, the identified periods were unstable, changing and fading, i.e., of little use for forecasting. It is acceptable to assume that their origin, "degeneration" and recovery, is caused by some external factors. We assume that solar activity and its multiyear fluctuations can be attributed to such factors.

This effect was shown in the work of A.L. Chizhevsky, where many connections of solar activity, expressed in the form of mean annual Wolf numbers, with meteorological, biological, and even social phenomena were described (Chizhevsky, 1973). Later these oscillations were analyzed by I.V. Maximov together with tidal oscillations in the World Ocean.

As a result of our studies, we obtained a number of statistically reliable links of solar activity with indicators of the state of ecosystems in the fishery areas of the World Ocean and with links of hydrological characteristics intermediate in these processes. Thus, we continued the list of cases of influence (effect) of solar activity on natural objects and systems proposed by L.A. Chizhevsky and a number of other researchers.

In these monographs, the physical mechanism of the impact of solar activity on the Earth's spheres is stated only hypothetically. It is precisely known its influence on the ionosphere, when flares on the Sun create a stream of positively charged ions (the so-called "solar wind"), which excites the ionosphere, worsening radio communication at short wavelengths. The question of energy impulse transfer to the troposphere remains open. M.I. Budyko points out the possibility of the effect of these processes on the solar constant, increasing it at high values (Budyko, 1974). There are hypotheses about a similar effect at its extreme values - at minima and maxima. Signs of the connection between solar activity and the solar constant are shown in the book by J.R. Herman and R.A. Goldberg (Herman and Goldberg, 1981).

The values of solar activity are represented by Wolf numbers, in the bicentennial series of which an 11-year periodicity stands out. It is not invariable, but varies from 8 to 12 years. In the above review by I. V. Maksimov (1970), information from various authors revealing more perennial rhythmic fluctuations - from 60-70 to 80 years is given.

One of such periods, namely, seventy-year periods, was found in the Earth's climate fluctuations, which was reflected in the changes in its rotation rate. This geophysical factor is presented in the works of N.S. Sidorenkov and P.I. Svirenko (Sidorenkov, 2004; Sidorenkov, Svirenko, 198). The physical essence of the process consists in the fact that during cooling and ice accumulation in the polar regions, the planet's rotation accelerates due to a decrease in the mass radius (momentum inertia), and during warming, the waters of melting glaciers spread in the World Ocean away from the rotation axis and the said velocity decreases. The authors of these works determined that the multiyear oscillations have a period close to 70 years. At that, its maximum was observed in the middle of the 30s of the last century. Consequently, its minimum falls on the middle of the 70s, and the next maximum on the period 2005-2010 (Sidorenkov, 2004).

To determine the mechanism of the obtained significant relationships, the predictor and predictant were correlated with hydrometeorological characteristics as intermediate links in the process of transferring the energy impulse of the initial factors through the natural conditions of increasing primary productivity and dynamic environmental prerequisites for the formation of aggregations of fishery objects. We used data of series of long-term temperature observations in coastal points reflecting climatic changes or intensity of upwelling development, as well as calculated indices of frequency and intensity of atmospheric transports over the water areas of fishing areas. Based on the peculiarities of atmospheric circulation over the studied water area, conclusions were made about strengthening or weakening of the flows in the known system of surface currents, which presumably could increase the intensity of topogenic eddies in the area of Antarctic krill fishery and also increase the expatriation of crustaceans from adjacent areas, which contributed to the success of their catch. Changes in the current system, for example, in the North Atlantic, determined changes in the state of ecosystems in the fishing areas, which influenced the yield and behavior of fishery objects.

The obtained series of relationships confirmed the possibility of using such an approach for fishery forecasting, since the values of both factors used have known periodic fluctuations and, therefore, can be extrapolated with a certain accuracy. For example, the methodology of extrapolation of Wolf numbers is presented in the work of M.N. Khramova et al. (Khramova et al., 2001), and the curve of the indicated 70-year periodicity of change in the Earth rotation rate can be continued after its maximum in the middle of the first decade of the XXI century.

Below we present the results of studies to identify the relationships of these geo- and heliophysical factors with catches of fish and Antarctic krill, as well as with the physical prerequisites for their yield and formation of aggregations, in a number of fishing areas. It should be noted that in none of the analyzed areas there was no case when a statistically reliable relationship with at least one of them or their combinations was not detected.

The objects of the long-term forecast were primarily fish catches in fishing areas in the form of: total catches of all or a particular species for the year or for the fishing season.

The largest number of predictants includes indicators of Antarctic krill: national catch and all countries - in the Atlantic and Indian Ocean parts of the Antarctic for a year or summer campaign, stock in the region and in commercial aggregations or stock in them. Indirect indicators of the abundance of fishery objects were taken as predicted characteristics of the biotic part of ecosystems: total annual catch of all prey products ("harvest" according to the definition of W. Royce (Royce, 1975), average annual specific biomass of phytoplankton; and also environmental indicators: water temperature, general thermal background of the water area and stressful phenomena for the ecosystem - areas and intensity of hypoxia and overcast phenomena on the northwestern shelf and in the Sea of Azov.

*The author expresses his sincere gratitude to his respected colleagues Dr. A.K. Vinogradov, Candidate of Biological Sciences V.N. Bolshakov, and Professor, Dr. G.G. Minicheva for their help and valuable advice in the preparation of the book. Special thanks to Dr. Yu.V. Bryantseva for technical support.*

## METHODOLOGY

As a heliophysical factor, we used the solar activity indicator in the form of Wolf numbers (W), as well as their anomalies W'= (|$W_i$ - **W**|). This number was determined by Wolf (quoted from I.V. Maximov (1970)) using the formula:

$$W = k\ (10\ g + f), \qquad (1)$$

where g is the number of observations of groups and individual spots counted in these groups and separately, f is the total number of spots counted in these groups and separately, *k* is a coefficient depending on the observer and his tube.

The series of Wolf numbers, having a duration of more than 200 years, was used in 1915-2012, since the longest series of observations of surface water temperature in the port of Odessa, correlated with the specified factor, starts from this year. In the process of determining the relationships of solar activity with various parameters, its average value was calculated not from the total sample, but within the studied parameter.

A number of values of the Earth's rotation velocity reflecting climatic changes (geophysical factor δ) is expressed in relative units from 0 to 1. Its maximum value (Sidorenkov and Svirenko, 1989)

was marked in the middle of the 30s of the last century, and the minimum in the middle of the 70s. According to some signs in the state of the Black Sea ecosystem, we marked a minimum in 1972 (Bryantsev, 2010).

Thus, we have an initial heliophysical factor with quasi-periodic changes with an average value of 11 years, more precisely 10.2 (Maximov, 1970), and its anomaly with a period of 5-6 years. When correlating these series with series of fishery, biotic and hydrometeorological characteristics, statistically significant relationships were obtained. The number and level of the latter increases if we take into account as a correction from the series of background 70-year periodicity within the above values from 0 to 1. This account is expressed in complex indicators formed by multiplying the Wolf numbers and its anomalies by the conditional values of the geophysical factor (δ). As a result, we obtain 5 indicators of the influence of primary factors external to ecosystems in the form of the following indices: W, δ, W', δW, δW'. Their values for the period I960-1992 are given in Table 1.

More often than not, three coefficients were sufficient to determine the overall linkage system:

mean field atmospheric pressure (assessment of cyclonicity or anticyclonicity):

$$A_{00} = \Sigma\Sigma P \cdot (X_m . Y_n) \cdot \Psi(Y_n) / \kappa \cdot 1;$$

of zonal transfer:

$$A_{01} = \Sigma\Sigma P (X_m \cdot Y_n) \ \Psi(Y_n) / \kappa \ \Sigma\Psi^2 (Y_n);$$

meridional transfer:

$$A_{10} = \Sigma\Sigma P \cdot (X_m \cdot Y_n) \cdot\varphi (X_m) / 1 \cdot \Sigma\varphi^2 (X_m),$$

where φ, ψ are Chebyshev polynomials, k is the number of nodes where the function is defined in the X direction, l is the number of nodes where the function is defined in the Y direction, $P(X\ Y_{mn})$ is the field of the function in the form of a square matrix (4 × 4).

An example of the baric field decomposition is given below. Calculation of the baric field expansion coefficients by Chebyshev polynomials. Polynomials for a variant of the field of 16 points (4 × 4).

*Table 1.* **Values of indicators of geo- and heliophysical factors for the period 1960-1992.**

| Year | W | δ | W' | δW | δW' | Year | W | Δ | W' | δW | δW' |
|---|---|---|---|---|---|---|---|---|---|---|---|
| 1960 | 112,3 | 0,34 | 48 | 38,1 | 16,3 | 1977 | 27,5 | 0,14 | 36 | 3,9 | 5 |
| 1961 | 53,9 | 0,31 | 10 | 16,7 | 3,1 | 1978 | 92,5 | 0,17 | 28 | 15,6 | 4,8 |
| 1962 | 37,5 | 0,29 | 26K | 11 | 7,5 | 1979 | 155,4 | 0,2 | 91 | 31 | 18,2 |
| 1963 | 27,9 | 0,26 | 36 | 7,3 | 9,4 | 1980 | 154,6 | 0,23 | 91 | 35,6 | 20,9 |
| 1964 | 10,2 | 0,23 | 54 | 2,3 | 12,4 | 1981 | 140,5 | 0,26 | 76 | 36,4 | 19,8 |
| 1965 | 15,1 | 0,2 | 49 | 3 | 9,8 | 1982 | 115,9 | 0,29 | 52 | 33,6 | 15,1 |
| 1966 | 47 | 0,17 | 17 | 8 | 2,9 | 1983 | 66,8 | 0,31 | 3 | 20,8 | 0,9 |
| 1967 | 93,8 | 0,14 | 30 | 13,2 | 4,2 | 1984 | 45,7 | 0,34 | 18 | 15,6 | 6,1 |
| 1968 | 105,9 | 0,11 | 42 | 11,7 | 4,6 | 1985 | 18 | 0,37 | 46 | 6,7 | 17 |
| 1969 | 105,5 | 0,09 | 42 | 9,5 | 3,8 | 1986 | 13,4 | 0,4 | 51 | 5,2 | 20,4 |
| 1970 | 104,5 | 0,06 | 40 | 6,2 | 2,4 | 1987 | 29,4 | 0,43 | 35 | 12,5 | 15 |
| 1971 | 66,6 | 0,03 | 3 | 2 | 0,1 | 1988 | 100,2 | 0,46 | 36 | 46 | 16,6 |
| 1972 | 68,9 | 0 | 5 | 0 | 0 | 1989 | 157,6 | 0,49 | 94 | 77,4 | 46,1 |
| 1973 | 38 | 0,03 | 26 | 1,1 | 0,8 | 1990 | 142,6 | 0,51 | 79 | 72,9 | 40,3 |
| 1974 | 34,5 | 0,06 | 30 | 2 | 1,8 | 1991 | 145,7 | 0,54 | 82 | 78,8 | 44,3 |
| 1975 | 15,5 | 0,09 | 48 | 1,4 | 4,3 | 1992 | 94,3 | 0,57 | 30 | 53,6 | 17,1 |
| 1976 | 12,6 | 0,11 | 51 | 1,4 | 5,6 | | | | | | |

Denotations : W - solar activity index (Wolf numbers); W' - anomalies of Wolf numbers (W' = W-W); δ - change in the Earth's rotation rate.

Ψ = φ = k - n+1/2 n = 4 k = 1,2,4

k, Ψ, φ - similarly:

1 - 2,5 = -1,5; 2 - 2,5 = -0,5; 3 - 2,5 = 0,5; 4 - 2,5 = 1,5.

Example calculation:

Field data (k = l = 4)

| Calculation results | Atmospheric pressure | | | | Calculation results | | |
|---|---|---|---|---|---|---|---|
| | | **(P mb - 1000)** | | | **∑P** | **Ψ** | **Ψ∑P** |
| | 16 | 19 | 21 | 24 | **80** | **-1,5** | **-120** |
| | 13 | 17 | 17 | 17 | **64** | **-0,5** | **-32** |
| | 12 | 12 | 16 | 17 | **57** | **0,5** | **28,5** |
| | 12 | 15 | 16 | 16 | **59** | **1,5** | **88,5** |
| **∑P** | **53** | **63** | **70** | **74** | **260** | **5** | **-35** |
| **φ** | **-1,5** | **-0,5** | **0,5** | **1,5** | | | |
| **φ ∑P** | **-75,5** | **-31,5** | **35** | **111** | | | |

$A_{00} = \Sigma\Sigma P\ (X_m, Y_n) \cdot \Psi(Y_n) / k \cdot l = 260/4 \cdot 4 = 16,25$

$A_{01} = \Sigma\Sigma P\ (X_m, Y_n) \cdot \Psi(Y_n) / k \cdot \Sigma\Psi^2\ (Y_n) = -35/4 \cdot 5 = -1,75$

$A_{10} = \Sigma\Sigma P\ (X_m, Y_n) \cdot \varphi(X_m) / l \cdot \Sigma\varphi^2\ (X_m) = 35/4 \cdot 5 = 1,75.$

# SOLVING LONG-TERM FORECASTING PROBLEMS IN SOME FISHERIES AREAS OF THE WORLD OCEAN

## 1. North Atlantic

In the study of the North Atlantic areas, we turned to retrospective data on fish catches in the period 1946-1985 (Marty and Martinsen 1969). The analyzed areas where significant correlation coefficients were obtained (significance level not more than 0.05) are given in Table 2.

This table presents the obtained significant relationships in 8 out of 10 analyzed areas, the relationships for the North Sea and Baltic Sea areas were not significant.

To determine the "meteorological" link in the system of energy impulse transmission to the biotic and fishery levels, we used the values of repeatability of six types of baric field diagnosed by the character of geographical localization of monthly atmospheric pressure anomalies given in the well-known monograph by K. V. Kondratovich (1977).

Type recurrences were compared with catch series, geo- and heliophysical indices. The results of correlation analysis yielded only two significant relationships of type "T" values (sum of $T_1$ and $T_2$ ) with solar activity and climatic factor $\delta$. The values of the coefficients are -0.485 and 0.442, respectively. The general scheme of the obtained relationships is shown in Fig. 1.

*Table 2.* **Correlations of fish catches in the North Atlantic (annual catches, thousand tons) with geo- and heliophysical factors**

| Neighborhood | Type, area, year | n | W | δ | δW | δW' |
|---|---|---|---|---|---|---|
| 1 | Catches of groundfish and benthic fishes in Icelandic and Greenlandic waters (1946-1965) | 20 | | -0,499 <0,05 | -0,572 <0,01 | |
| 2 | Cod catches in the areas of the Wyvila-Thomson Threshold and Greenland (1954-1965) | 12 | | 0,843 <0,01 | | 0,576 0,05 |
| 3 | Sea bass catches in Area 2 (1953-1965) | 13 | | | -0,579 <0,05 | |
| 4 | Fish catches in the Newfoundland, Labrador, N. England and N . Scotland areas (1954-1965) | 12 | | -0,961 <0,01 | | -0,606 <0,05 |
| 5 | Cod catches in the western sector | 14 | | -0,905 | | -0,716 |

| | | | | | | |
|---|---|---|---|---|---|---|
| | of the North Atlantic (1952-1965) | | | <0,01 | | <0,01 |
| 6 | Herring hake catches in the western sector of the North Atlantic (1954-1965) | 12 | | -0,795<br><0,01 | | |
| 7 | Herring catches in the western sector of the North Atlantic (1956-1965) | 10 | -0,794<br><0,01 | -0,766<br><0,01 | -0,708<br><0,01 | |
| 8 | Fish catches in the Danish Straits and the English Channel (1953-1965) | 13 | | -0,806<br><0,01 | | |

Note . Data for the year (thousand tons) with geo- and heliophysical factors (Marty and Martinsen, 1969). The designations are given in the text.

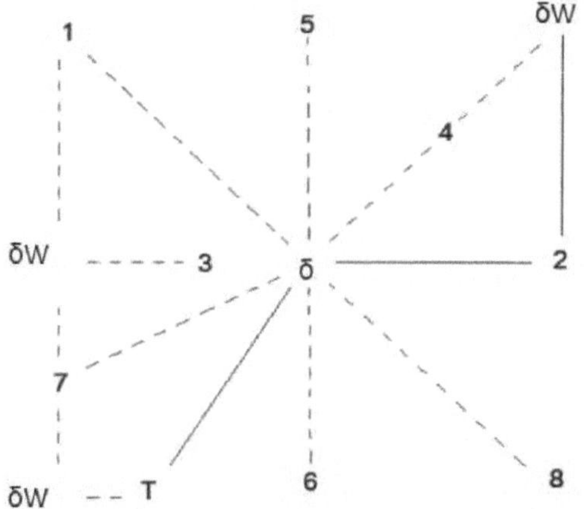


Fig. 1. Scheme of relationships between fish catches (see Table 2 for district numbers), geo- and heliophysical factors (designations in the text) and repeatability of the baric field type "T" (values presented in K. V. Kondratovich, 1997). Dashed lines indicate direct relations, solid lines - inverse relations.

According to this scheme, fish catch rates in North Atlantic areas, with the exception of cod catches at the Wyvil-Thomson Threshold (Area 2, see Table 2), are negatively related to a climatic (T) characteristic reflected by the Earth's rotation rate index ($\delta$). This characteristic is also inversely related to the values of series 3 (sea bass catches in area 2) through the complex index $\delta W$, where the influence of the first factor may be dominant. Consequently, the analyzed period of fishing in the North Atlantic waters falls at the time of decrease of the climatic indicator $\delta$, i.e., at the descending branch of the above-mentioned 70-year cycle. At the same time, in the general description of global climate changes (Budyko, 1974), this period is characterized by cooling.

Let us further consider the direct connection of the $\delta$ index with the repeatability of the baric field of the "T" type. In spite of the "opposite" of the components of this type (T and T ) indicated by K. V. Kondratovich indicated by K. V. Kondratovich, the "opposite" of the components of this type ($T_1$ and $T_2$ ), consisting in the fact that the first of them is characterized by active cyclonic activity in the area of Iceland and the second - by its displacement to the Barents Sea (Kondratovich, 1977) to the south of Iceland and Greenland, the western and southwestern atmospheric transports, northern and northwestern in the areas of Newfoundland and the New England and New Scotland shelves are intensified. Consequently, at the epoch of decreasing value of the climatic index δ, there is, in general, a weakening of the inflow of warm waters of the North Atlantic Current in the eastern part of the North Atlantic and cold waters of the Labrador Current - in the western part.

The favorable effect of decreasing solar activity is manifested in relation to herring catches in the western sector of the North Atlantic directly and in complex with the δ indicator. Therefore, herring catches increase with decreasing T-type recurrence (see Table 2, Fig. 1).

In conclusion, it should be noted that there are significant relationships of the given geo- and heliophysical factors as climatic indicators with fish catches in the North Atlantic fishing areas, which can be the basis for regional methodology of fishery forecasts due to the possibility of extrapolation of series of their values.

The ecological effect of the success of fishing in the analyzed areas of the North Atlantic can be established by analyzing the scheme of links between catches and hydrometeorological, geo- and heliophysical characteristics. In this case, it is necessary to take into account also the connections of the above-mentioned indicators with the general system of atmospheric circulation in this region, which we have revealed.

## 2. Barents Sea

The volume of studies conducted in the Barents Sea allows us to present the long-term changes in a number of properties of its ecosystems in a complex. For example, in the work of Yu.Yu. Matishov et al. (2010) an important conclusion is made that the thermal background of the sea determines the productivity of various fishery objects, and also lists the found links with the average annual temperature of the Kola section and its anomalies with the benthos biomass, with the survival of the omnivore - the Kamchatka crab and with the degree of cod migration to the east and north of the sea. In the latter case, we have an example of dependence on climatic fluctuations, indirectly - the fishery indicator. Thus, the forecast of the thermal background, conditioned by the intensity of the inflow of the Gulf Stream system waters, would make it possible to forecast all the above characteristics.

When comparing the data from this paper with our climate indices, the coincidence of the minimum and maximum of the δ series with the lowest and highest spread of cod to the east and north of the Barents Sea in the late 1970s and 2004-2006, respectively, becomes immediately apparent. Our correlation analysis of a number of characteristics from Matishov et al. (2010) with the above-mentioned geophysical index revealed its significant relationship with the value of the weighted average anomaly of water temperature at the "Kola Meridian" in the 0-200 m layer. The correlation coefficient is equal to 0.842 and the significance level is less than 0.01, despite the small values of the compared series (2001-2009).

The mechanism of the natural process, the beginning of which is the energy impulse of climatic fluctuations, and the final link is the reaction of the biotic part of the marine ecosystem and the level of fishing success, can be represented by including the "atmospheric" link and its subsequent effects on the system of currents in the studied region. In this study, we took the values of recurrence of six types of baric field in the North Atlantic, diagnosed by the character of geographical localization of monthly atmospheric pressure anomalies from the monograph by K. V. Kondratovich (1977).

A significant correlation with the δ index (coefficient 0.442, significance level 0.05) was obtained when comparing its series with the total value of the "u" type anomalies ($u_1$ plus $u_2$ ). The first of them is characterized by active cyclonic activity in the vicinity of Iceland and the second - by its displacement to the Barents Sea. As a result, westerly and southwesterly atmospheric transports intensify south of Iceland and Greenland (Kondratovich, 1977). Thus, during the epoch of the climatic indicator δ increase, the inflow of warm waters of the North Atlantic Current intensifies, which is reflected by a number of the above-mentioned weighted average temperature anomalies in the 0-200 m layer (Matishov et al., 2010) and subsequent manifestations of this influence on the state of the Barents Sea ecosystems.

For approximate multi-year forecasts of the thermal background level (T) and related biotic and fishery indicators of the Barents Sea, we consider it acceptable to use extrapolated values of the climatic index δ in Eq:

$$T = 0{,}767\ \delta - 0{,}026. \qquad (2)$$

It can only show the general trend in multi-year changes in fishing efficiency. However, in the practice of forecasting catches of specific fisheries, it is acceptable to "coarsen" the forecast. For this purpose, the correlated series of previous values can be divided into three equally probable ranges (VanderWarden, 1960), then a matrix of frequency correspondence can be compiled and used to determine the probability of the level of the predicted characteristic at certain ranges of the predictant.

## 3. Patagonian Shelf

The Falklands-Patagonia Fishing Area (No. 41 on the FAO map) is located between 41-45° S and from the coast of South America to 55° W. The fisheries are: southern putasu, Argentinean merlin, American and Patagonian macronuts, and Argentinean anchovy. The fishery targets: southern putasu, Argentine merluza, American and Patagonian macronuts, and Argentine anchovy. Squid represent a significant part of the catches. The possible annual catch of fish is 1230 thousand tons (Fisheries..., 1984). The total biomass of harvested marine product per year, or "harvest" as defined by Royce (1975) reaches 1800 thousand tons. Consequently, a multi-year forecast of fishery productivity is essential for fishing organizations and firms. Therefore, a method of indicative fishery forecast using geo- and heliophysical factors is proposed for this area (Bryantsev, 2012).

FAO statistical data for the period 1950-2008 were taken as a predictor to develop a methodology for predicting the annual fish and seafood "yield" in the Patagonian shelf area (U) with a multi-year advance. The series of annual solar activity (W) and Earth rotation rate (δ) were used as predictors. In order to reveal the mechanism of impulse transfer of primary factors' impact on the hydrostructure and current field of the studied water area, determining the production in the ecosystem, we considered the links of catches with the peculiarities of atmospheric circulation and the latter - with the mentioned factors. These features are expressed by us in the form of atmospheric transports, which were calculated by decomposing the surface baric field within the area into a series by Chebyshev polynomials. The calculation methodology is given in the monograph by K.I. Kudryava et al. (1974). A 16-point field was selected with approximately equal distances between them and taking into account the meridian convergence (Table 3).

*Table 3:* **Coordinates of points of the standard baric field**

| Ψ°S. | λ° s.d. | | | |
|---|---|---|---|---|
| 40 | 73 | 60 | 47 | 74 |
| 50 | 76 | 60 | 44 | 29 |
| 60 | 80 | 60 | 40 | 20 |

| 70 | 89 | 60 | 31 | 8 |
|---|---|---|---|---|

The values of decomposition coefficients with annual and semiannual averaging were used in the analysis: $A_{00\text{-}1}$ - mean atmospheric pressure for the first half-year and $A_{00\text{-}г}$ - similar for a year.

The designated series were correlated with each other. The values of correlation coefficients were taken with the level of significance equal to and less than 0.05.

The results of correlation analysis of the series are presented in Table 4.

*Table 4:* **Correlation coefficients and significance levels for correlation of series of geo- and heliophysical factors, atmospheric circulation indices and total commercial catches in the Patagonian Shelf area** (notations in text)

| | δ | δW' |
|---|---|---|
| У | 0,649 (< 0,01) | 0,384 (< 0,01) |
| $A_{00\text{-}р}$ | – | -0,379 (< 0,05) |
| $A_{00\text{-}1}$ | – | -0,575 (< 0,01) |

The obtained system shows a direct connection of catches with the Earth rotation speed and solar activity anomaly in complex with the first factor. With this value in the inverse relationship there are indicators of mean atmospheric pressure within a year and, with even greater correlation coefficient, averaged for the first half of the year, the time of fishing. Thus, the pattern of relationships physically reflects its success at cyclonic circulation and, consequently, at southern atmospheric transports in the western half of the analyzed baric field. Such wind influence enhances the Falkland Current and topogenic eddies within the shelf and in the interaction area with the waters of the western periphery of the Brazil Current.

The equation of the relationship between the total catch and the index δ, reflecting climatic fluctuations, is as follows:

$$У = 1231{,}5\delta + 134{,}4, \qquad (3)$$

where U - total catch in thousand tons; factor δ - in fractions of one.

The equation can be used for tentative catch forecasts, since the climatic factor can be extrapolated for any number of years, having in mind a certain 70-year periodicity of its fluctuations (Sidorenkov, 2004). Provision of the equation is not high, however, with acceptable accuracy of the forecast it can be expressed by means of a matrix of ratio of repeatability of values of equally probable ranges into which the characteristics are divided, with their designation in the form: H - low values, C - medium and B - high. The corresponding catch values are given in Table 5.

*Table 5.* **Ranges of values of total catch (U, thousand tons) and climatic index (δ)**

| Symbol | Range | | |
|---|---|---|---|
| | H | C | B |
| δ | < 0,33 | 0,33-0,66 | > 0,66 |
| y | < 0,356 | 356-1097 | > 1097 |

The calculated recurrence correspondence matrix is presented in Table 6. In it we have informative cells with zero repeatability of the combination of HB and BH, which means that there are no cases of high catches at low values of δ index and their low values at high δ. In addition, at high δ values, high catches occur 85% of the time and high and medium catches occur 100% of the time. At low δ values, the probability of low and medium catches also reaches 100%.

Thus, the main determinant of changes in fishing success (total fish and squid catch) is autocomposition of climate change with a 70-year period, indirectly reflected by the Earth's rotation rate (index δ).

*Table 6:* **Correlation matrix of repeatability of catch values and index δ**

| y | δ | | |
|---|---|---|---|
| | H | C | B |
| H | 14 (0,61) | 11 (0,48) | 0 |
| C | 9 (0,39) | 7 (0,30) | 2 (0,15) |
| B | 0 | 5 (0,22) | 11 (0,85) |
| Σ | 23 | 23 | 13 |

Additionally, the influence of the anomaly (differences from the mean) of solar activity (W'), which determines the fluctuations with a 6-year period, is also manifested.

The maximum of the mentioned main factor falls on the mid-30s, and the minimum - on the mid-70s (Sidorenkov, 1989). Thus, the end of the analyzed series of catches almost coincides with its next maximum. Approximately in the period close to the maximum δ (1991-2008) the highest catches were observed in the Patagonian shelf area. Consequently, fishing here will be successful until 2018, with a subsequent decline by 2042.

## 4. Atlantic part of Antarctica

Multiyear fluctuations of Antarctic krill catches in the Atlantic part of the Antarctic (ACA) are caused by changes in the structure of currents, which determines the transport of crustaceans to fishing areas and the intensity of topographic cycles that create conditions for the formation of fishing aggregations. Such effect is determinant for the state of ecosystems in fishery areas of the World Ocean, for example, in the Black Sea (Bryantsev, 2001, 2010), in the eastern and western parts of the Pacific Ocean and the Southeast Atlantic (Cushing, 1995). In the last of the above 23 papers, the author summarizes the data of other researchers, in which sardine and anchovy catches are associated with their yields. Multiyear differences of the latter are determined by the reorganization of macroscale current systems under the influence of certain features of atmospheric circulation (AC) due to global climatic changes.

In confirmation of such a scheme of dependencies, significant correlations of fish catches with the average temperature of the air and surface layer of the World Ocean, with the inflow of solar energy, with atmospheric transports of certain directions in the northern hemisphere are given.

These relationships, however, cannot be used for multiyear forecasts because the listed predictors are not predicted due to the small memory of the ocean, much less the atmosphere.

The object of the forecast is the catches of Antarctic krill in three areas of the ASA: off the South Shetland and South Orkney Islands and off South Georgia Island, as well as the total catch (FAO data). These figures are assumed to reflect krill yield, but more so the intensity of transport from the Weddell and Belinshausen Seas, and the characteristics of the local current system. As noted (Kramer 1975), integral indices are more effective for determining statistical relationships than detailed time and area data.

To establish the relationship between the success of krill fishery in the ACA and atmospheric transports, we used the series of total krill catches in the listed areas with the designations 48.1, 48.2 and 48.3, respectively, for the period 1982-2008, and the total catch in the ACA (area 48) in 1977-2008 (in thousand tons). External factors are represented by the values of baric field decomposition coefficients by Chebyshev polynomials: $A_{00}$ , $A_{01}$ , $A_{10}$ , averaged for the first and second half of the year and for the year as a whole. The standard baric (surface) field was determined within the range 40- 70° S, 2-73° W. E. Its 16 points were located in grid nodes (4 × 4) with approximately equal distances between them (taking into account meridian convergence). The coefficients were calculated according to the methodology mentioned above (Kudryavaya et al., 1974). The first of them reflects the mean pressure value (predominance of cyclonicity or anticyclonicity), the other 4 - the intensity of elementary atmospheric transports (zonal and meridional).

The indices of the "initial" factors are presented in the form of series: solar activity (W), its anomaly modulus (W' = W - Wsr. polynomial), the Earth's rotation rate (δ - in relative units, from 1 at the maximum to 0 at the minimum), with periodicity, where the first falls in the mid-30s, the second in the mid-70s), as well as their products (δW and δW').

Annual krill catch data were summarized for the second half of the previous year and the first half of the following year. The resulting inaccuracy of comparison was compensated for by the fact that the values of geo- and heliophysical characteristics were a continuous function. In addition, the relationships were checked by correlations with a year shift.

The correlation coefficients obtained by comparing the series are presented in Table 7.

*Table 7.* **Correlation coefficients of the relationships of geo- and geophysical characteristics with the indices of atmospheric transport and krill catches in the Antarctic Atlantic** (notations in the text)

| Functions | Arguments | | | | | |
|---|---|---|---|---|---|---|
| | W | δ | W' | δW' | A10-2 | A10-p |
| A00-1 | - | -0,406 | -0,369 | -0,594 | - | - |
| Aaargh! | - | -0,359 | - | -0,401 | - | - |
| A01-1 | - | - | -0,374 | - | - | - |
| U, NN districts | | | | | | |
| 48.1 | 0,389 | - | - | - | - | - |
| 48.2 | - | -0,565 | - | - | -0,516 | -0,444 |
| 48.3 | - | -0,375 | - | - | - | - |
| 48 | 0,531 | -0,498 | 0,505 | - | - | - |

The obtained correlations allow us to draw a number of direct and indirect conclusions about the dependence of krill catches in different areas and the ASA as a whole on certain atmospheric transports (see Table 7). Thus, the system of direct relations of solar activity and its anomaly with the total catch in the ACA and in area 48.1 and the inverse relation of the anomaly with the zonal transport allows us to infer an inverse relation with the specified transport of the catch in the first area. Further, since the positive value of $A_{01}$ in the southern hemisphere denotes an east-west transfer, the inverse relationship with it denotes a positive effect of **westward transfer** on the success of the krill fishery in the South Shetland Islands area.

Here they are inversely related to the meridional transfer (in the second half of the year and for the year as a whole) from **south to north**. The inverse relationship of these catches with the Earth rotation rate, which in turn is inversely related to the **anticyclonicity** indicator within the analyzed area ($A_{00}$ ), confirms this relationship, since the transport from the south is realized in the eastern part of the anticyclonic vortex (in the southern hemisphere).

A similar system of relationships and catch values in area 48.3. Here, the relationship with the Earth's rotation speed and anticyclonicity may denote their direct relationship with the **southward and eastward transport.**

Thus, the anticyclonic form of atmospheric transports within the analyzed field creates a system of water circulation contributing to the formation of commercial accumulations of Antarctic krill.

In accordance with the logic of the constructed dependence scheme, the main indicators in the "mechanism" determining interannual fluctuations in krill catches are the solar activity anomaly and the Earth rotation rate. As we can see, the most common predictant, - total krill catch in the ACA, has the best correlations with both initial factors. The multiple correlation coefficient is 0.827. Regression equation:

$$У = 202{,}52 - 177{,}141\delta + 2{,}297\ W' \qquad (4)$$

allows us to make a multi-year forecast of the catch (Y) in the ACA when the values of the arguments are obtained by extrapolating them for any period. The general principle of the forecast is based on the fact that the largest krill catches in the ACA occur at values of solar activity close to the extremes (the largest differences from the average W) and during epochs with low Earth rotation speed (δ). Both factors determine the peculiarities of atmospheric circulation during these periods, the system of currents in the ACA, the direction and intensity of water transports, and the degree of whirlwinds that provide the accumulation effect and formation of commercial krill aggregations. According to the nature of the named features, they produce fluctuations in fishing conditions with 6- (W') and 70-year (δ) periodicities.

In conclusion, we note that the multiyear series of values of solar activity and the Earth rotation rate, indirectly reflecting climatic fluctuations, are significantly correlated with the catches of Antarctic krill in the three commercial areas of the Atlantic part of the Antarctic and with the peculiarities of atmospheric circulation in the region. It is supposed that the predominance of the anticyclonic component here determines the macroscale system of currents favorable for transport and formation of commercial accumulations of crustaceans.

The fragmentary scheme of correlations is reduced, however, to a simple consequence - a positive influence of the strengthening of the Antarctic Circumpolar Current and its northeastern branch.

Consequently, the possibility of extrapolating these initial geo- and heliophysical characteristics provides a basis for creating a methodology for multi-year forecasting of fishing success.

## 5. Indian Ocean part of the Antarctic (Commonwealth Sea)

Multiyear fluctuations of Antarctic krill (*Eufausia superba* Dana) stocks in the Commonwealth Sea are very significant for the enterprises fishing in the Antarctic, since, according to statistics, its abundance during several years is replaced by years of almost complete absence of accumulations suitable for rational fishing. It was also established that the accumulation of crustaceans in the indicated sea and the adjacent open part of the Indian Ocean is conditioned not so much by their yield, as by the intensity of transport by currents and whirling of the latter in quasi-stationary topographic and synoptic eddies (Maslennikov, 2003). Therefore, our stock assessment is performed only for the traditional fishing waters in the designated area.

The links between some atmospheric circulation parameters, indirectly reflecting changes in the structure of the current field, and krill stocks obtained in the marine expeditions of the South Research Institute of Marine Research and Development were investigated. The characteristics of the atmospheric field were calculated from the data of the surface atmospheric pressure archive.

Statistical comparisons of series can be used for operational and monthly forecasts. The atmospheric circulation in the southwestern Indian Ocean was analyzed on the basis of atmospheric pressure data within a 16-point standard field between 50 and 65°S and 45-75°E. The surface pressure survey points were labeled taking into account the meridian convergence and the distances between them in latitude of 300 miles and longitude of 382 miles (Fig. 2).

The first five coefficients ($A_{00}$ , $A_{01}$ , $A_{02}$ , $A_{10}$ and $A_{20}$ ) were calculated, reflecting, respectively: field mean, zonal transport, zonal in the northern and southern part of the field, meridional, meridional in the western and eastern part of the field.

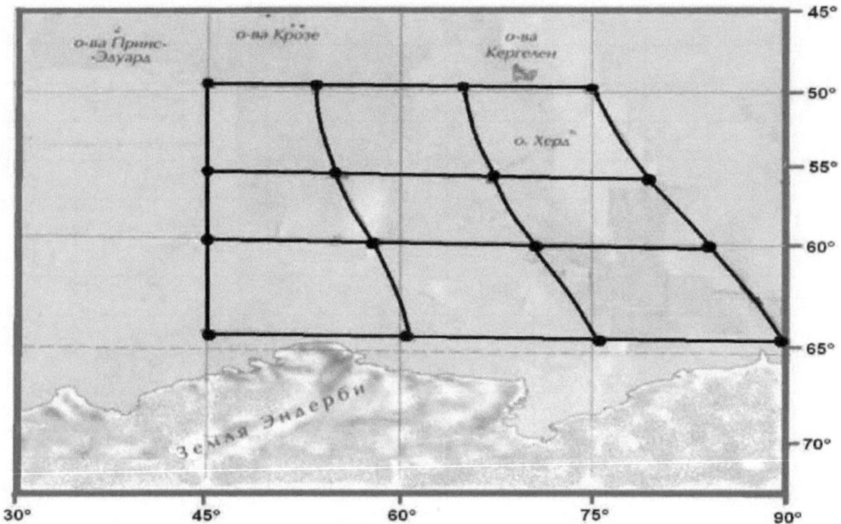


Fig. 2. Standard grid for calculating the coefficients of the surface baric field decomposition into series by Chebyshev polynomials

Thus, the next maximum is in 2007. The products of δW and δW values were also used' . The stock of Antarctic krill was determined in the Commonwealth Sea and in the adjacent part of the open ocean in millions of tons for the period 1977-1990 based on the data of surveys by the South Research Institute of Oceanography (Samyshev, 1991; Bibik, Yakovlev, 1990).

The correlation analysis resulted in the relationships presented in Table 8 and Fig. 3. 3.

The mean atmospheric pressure within the designated water area ($A_{00}$ ), at temporal averaging of the period: summer, autumn and subsequent spring, is positively related to all elements of the "pentagram" factors (W, W', δW and δW' ), excluding the Earth rotation speed (δ), but with signs of its influence in the complexes δW and δW\ The listed series represent the time of the navigation period, i.e. January, February, March, April, November and December.

*Table 8.* **Correlation coefficients and significance levels of relationships between the following indicators: atmospheric circulation, solar activity, Earth rotation rate (and their combinations) and stocks (3) of Antarctic krill in the Commonwealth Sea ($W_{м}$ ) and adjacent water areas of the Indian Ocean ($W_{ок}$ )**

| Functions | Arguments | | | | | | | |
|---|---|---|---|---|---|---|---|---|
| | W | δ | W | δW | δw' | $A_{00}$ | $A_{10}$ | $A_{02}$ |
| $A_{00}$ | 0,718* | - | 0,471 | 0,600* | 0,591 | - | - | 0,590 |

| 1975-92 | < 0,01 | - | < 0,05 | < 0,01 | <0,01 | - | - | < 0,01 |
|---|---|---|---|---|---|---|---|---|
| Ao2 | 0,462* | - | - | - | - | - | - | |
| | < 0,05 | - | - | - | - | - | - | |
| Aio | - | - | - | - | - | - | 0,693 | |
| | - | - | - | - | - | - | < 0,01 | |
| Aθι-ιι | 0,528 | - | 0,529 | - | - | - | - | |
| | < 0,05 | - | <0,05 | - | - | - | - | |
| 3, NN districts. | | | | | | | | |
| 3м | -0,336 | -0,471 | - | - | - | -0,528 | -0,663 | |
| 1977 - 1990 | > 0,05** | > 0,05** | - | - | - | 0,05 | < 0,01 | |
| 3 | - | - | - | - | - | - | | 0,544* |
| ʲOK | - | - | - | - | - | - | | < 0,05 |

* Correlation coefficients at a 1-year shift (see Fig. 1 for notations).

** Coefficients not significant but used in the calculation of the multiple correlation coefficient. See Chapter 1 for other notations.

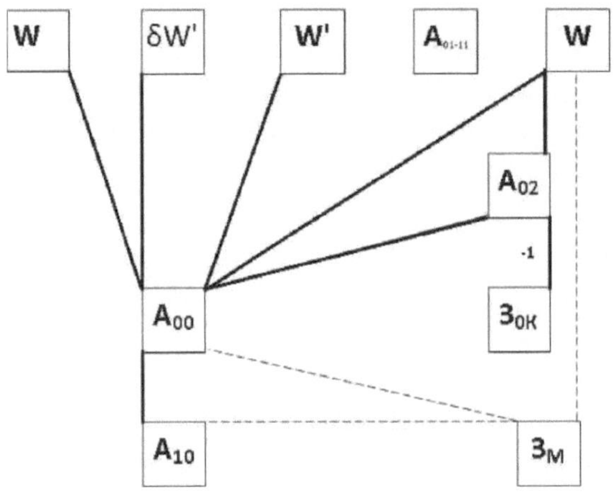


Fig. 3. Graph of relationships of Antarctic krill stocks in the Commonwealth Sea area with the indicators of atmospheric circulation, geo- and heliophysical factors (see Section 1 for notations). Solid lines indicate direct relationships, dashed lines indicate inverse relationships, -1 - shift by 1 year

Therefore, the correlation was performed synchronously and with a shift of 1 year. As a rule, the coefficients kept the same sign and in some cases were higher than the first ones. This is due to the monotonicity of the functions of geo- and heliophysical factors, where the preceding values differ little from the subsequent ones. Coefficients with higher significance were included in the generalization matrix. In Fig. 3, the relations with the shift are marked with minus 1.

The correlations of mean pressure with atmospheric transports $A_{10}$ and $A_{02}$ , in spite of a certain conventionality (calculation of these values from data of one field, but with different polynomials), are taken into account. At high pressure, transports from the north will prevail, as well as from the east in the southern part of the field and from the west in the northern part. All three characteristics are inversely related to krill stocks in the Commonwealth Sea and in the "ocean" area. Thus, transport directions opposite to those indicated, i.e., **southward, and westward in the southern part of the field and eastward in the northern part of the field**, would be favorable for increasing crustacean stocks. Both areas are located in the southern part of the analyzed field, hence one can assume that the direction of these transports here contributes to strengthening of the southern branch of the Antarctic Circumpolar Current (ACC) and weakening of the West Coastal Current. The southern transport together with the anticyclonic atmospheric vortex conditionally formed by the indicated favorable transports determine the expatriation of krill from the coast to the fishing areas and the increased vorticity of the current field creating a collecting effect for them.

An additional illustration of the presented system is a direct connection of zonal (western) transport in February with the solar activity factors and its extreme values (W').

Thus, calculation of mean atmospheric pressure within the selected standard field, zonal and meridional transports can serve as one of the arguments for operational forecasting of fishing success, having in mind their inverse relationship.

For multi-year forecasts, the mediated relationships between solar activity and Earth's rotation rate (with their extrapolation capabilities) with Commonwealth Sea krill stocks should be used. The linkage equation derived from the multiple regression analysis model is as follows:

$$З_{М} = 15{,}8 - 0{,}03W - 21{,}2\delta \qquad (5)$$

(R = 0.679, 86% equation security).

It is also a tentative prediction for the oceanic part of the fishing area, as there is a significant direct correlation between the stocks in both parts ($r = 0.750$, $p < 0.01$).

Thus, it can be concluded that a decrease in the mean annual values of solar activity and the Earth's rotation speed and atmospheric transports directly depending on them ($A_{00}$ , $A_{10}$ and $A_{02}$ ) leads to an increase of Antarctic krill stocks in the Commonwealth Sea and adjacent water areas of the open part of the Indian Ocean. The weakening of these transports and strengthening of the opposite ones contributes to the transport of crustaceans to the fishing areas and to the increase in the eddies of the current field, which determines the formation of their dense aggregations.

Assuming an indirect dependence of krill stocks on the above factors, despite the fact that the coupling equation is derived for a limited period, we can construct a conditional plot of multi-year changes in krill stocks in the Commonwealth Sea region using the data and extrapolated argument values up to 2013 (Fig. 4).

The negative values in the prognostic equation are due to the high values of the Earth's rotation velocity at the epoch of their maximum. This may explain the period of absence of commercial krill accumulations (as well as fishing) after 1990. Judging from the graph, signs of a steady increase in krill stocks appear after 2010, and opportunities for effective fishing appear in another 3-4 years.

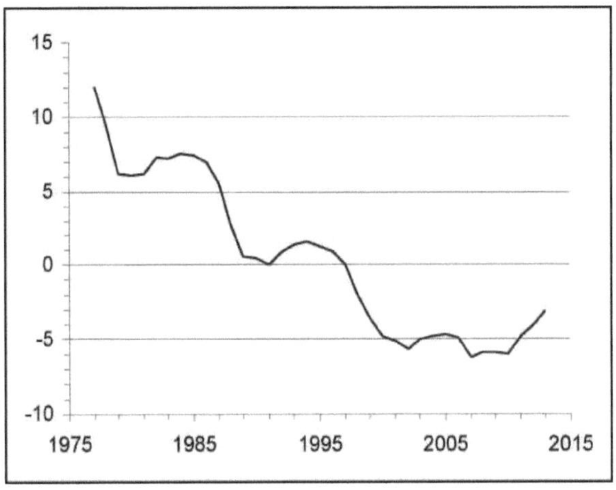


Figure 4. Conditional plot of changes in Antarctic krill stocks in the Commonwealth Sea

## 6. Southeast Pacific

In the literature on the yield and behavior of Peruvian jack mackerel, the main commercial species outside the 200-mile economic zone in the Southeast Pacific Ocean (SEP), empirical relationships are widely presented: direct to oceanographic factors and indirect to meteorological factors. Some of these are given in the Fisheries Description of the Region

(Fisheries..., 1985). More broadly, such relationships are discussed in a synthesis by Cushing (Cushing, 1995). It compiles information from a number of authors on the relationship of annual catches of sardines and anchovies in the Pacific and Atlantic Oceans to global changes in climatic characteristics. Increases and decreases in catches of these species are linked to: annual mean values of air and ocean surface temperature in the northern hemisphere; the amount of solar radiation reaching the Earth's surface; and atmospheric circulation patterns. The last of these factors determines changes in the system of currents, hence in the intensity and location of macroscale cycles. Depending on their displacement, the intensity of coastal upwelling changes, creating the inflow of biogenic salts into the photic layer and the level of primary production, which determines fish yield.

However, the obtained relationships cannot be applied for long-term (a year or more) fishery forecasting, since the above-mentioned determinants are not predicted with such an advance. Attempts are made to link changes in current fields with the El Niño process and the Southern Oscillation (ENYC), but their auto-oscillating nature with an unstable period does not allow using the dependencies for forecasting.

Our approach is based on the use of a number of geo- and heliophysical factors, in particular solar activity and changes in the Earth's rotation rate, which are extrapolatable and, on the other hand, can be related to changes in atmospheric circulation and large-scale current fields, as well as to yields and fishing behavior. In finding such relationships, we can construct a dependency scheme and create a series of regression equations for forecasts of Peruvian jack mackerel catches in the SEE (Bryantsev, 2009a).

As an analog of the yield and formation of commercial aggregations of Peruvian jack mackerel, its catches in the SEE by all countries in 1985-2005 were taken (Budyko, 1974). This approach was also used by Cushing (1995) to indirectly estimate the stock of different fish species.

When comparing with the above series, the adopted characteristics were used as "initial" factors. The "second level" factors - the values of atmospheric transports - were calculated by decomposing the standard baric field into a series by Chebyshev polynomials. It includes the investigated water area of the Antarctic Circumpolar Current and the connected system of the Peruvian Current. On the oceanic side of the latter, commercial accumulations of mackerel are formed in macroscale cycles.

In the subsequent analysis, only meridional transfers proved to be sufficient:

$$A_{10} = \sum_{1}^{k}\sum_{1}^{L} (XmYn)\varphi_1(Xm)/L\sum_{1}^{k} \varphi_1^2(Xm); \quad (6)$$

$$A_{20} = \sum_{1}^{k}\sum_{1}^{L} (XmYn)\varphi_2(Xm)/L\sum_{1}^{k} \varphi_2^2(Xm), \quad (7),$$

where k, L are the number of field nodes, respectively, along the X (parallel) and Y (meridian) axis, Xm, Yn are the values of atmospheric pressure (minus 1000 mb) at the nodes of the adopted grid, $\varphi_1$ and $\varphi_2$ are the first and second expansion polynomials.

In the southern hemisphere, a positive value of $A_{10}$ reflects transport from the north, while $A_{20}$ ▪ similar transport in the western half of the field.

The series were compared by the method of pair and multiple correlation (Brooks and Caruthers, 1977). Coefficients with a significance level not exceeding 0.05 were included in the correlation matrix.

In the practice of fisheries forecasting, it can be sufficient to represent the predictor in the form of three ranges, for example, as three equally likely ranges: low value (H), medium (C) and high (H) (Brooks and Caruthers, 1977). The matrix of matching frequencies of these values shows the level of coincidence, the significance of which is assessed using Pearson's $\chi$ index$^2$ null hypotheses (Van der Varden, 1960):

$$\chi^2 = \Sigma\ (Xi - pn_i)/pn_i + \Sigma\ (Yi - qn_i)/qn_i, \quad (8)$$

where $p = \Sigma Xi/\Sigma n,\ q = \Sigma Yi/\Sigma n,\ n -$ is the number of members of the compared series X and Y.

Connections of solar activity and its anomalies with peculiarities of atmospheric circulation, or, directly, with fish catches, are considered without physical explanations, according to the "black box" principle, although we find signs of such connections in the literature, starting from the monograph of A. L. Chizhevsky (1973). L. Chizhevsky (1973).

As shown by the correlation comparison of the analyzed characteristics, the catch of Peruvian mackerel in the SEE is in direct relation with the solar activity anomaly module and in inverse relation with the meridional atmospheric transports ($A_{10}$ and $A_{20}$ ) calculated by equations (4 and 5) separately for the first and second half of the year (see Table 8).

Table 8 additionally shows a coefficient slightly below the boundary level of significance (-0.415, critical 0.433), but both relationships of the extrapolated factors with mackerel catches (St, million tons) yield a high multiple correlation coefficient of 0.749 and a regression equation used for the long-term forecast:

$$St = 3{,}36 + 0{,}016W' - 1{,}8\delta \qquad (9)$$

Its endowment is 52%, but the fishery forecast can also be represented as three equally likely ranges (Table 9).

*Table 9:* **Correlation indices of Peruvian mackerel catches in the SEE with external factors in the period 1985-2005.**

| y | Factors | | | | | | | |
|---|---|---|---|---|---|---|---|---|
| | W' | δ | $A_{10-1}$ | $A_{10-2}$ | $A_{10-р}$ | $A_{20-1}$ | $A_{20-2}$ | $A_{-год}$ |
| Correlation coefficient | 0,453 | -0,415* | -0,504 | -0,611 | -0,588 | -0,438 | -0,592 | -0,533 |
| Levels of significance | <0,05 | >0,05 | <0,05 | <0,01 | <0,01 | <0,05 | <0,01 | <0,01 |

Let us consider the predictive power of both factors when "coarsening" the predictant to a binary score: medium and high values (C and B) and below 2.64 million tons (H) (Tables 10 and 11), applying the 2 significance assessment method for the frequency matching criterion - Pearson's $\chi^2$ .

*Table 10.* **Equal probability ranges of characteristics of annual catches of Peruvian mackerel in the SEE (St, mln t) and its determinants**

| Values | Rows | | |
|---|---|---|---|
| | H | C | B |
| St | < 2,64 | 2,64-3,93 | > 3,93 |
| W' | < 39 | 39-56 | > 56 |
| δ | < 0,58 | 0,58-0,78 | > 0,78 |

*Table 11.* **Matrix of comparison of frequencies of solar activity anomaly module values and catches of Peruvian jack mackerel in the SEEO**

| St | W' | | | |
|---|---|---|---|---|
| | H | C | B | Σ |
| CB | 3 0,38 | 4 0,50 | 4 0,80 | 11 |
| H | 5 0,62 | 4 0,50 | 1 0,20 | 10 |
| Σ | 8 | 8 | 5 | 21 |

The null hypothesis is accepted in the first case (Table 10), where $\chi^2 = 2.258$ with a significance level > 0.10, and rejected in the second case, where $\chi^2 =$

9.261 with a significance level < 0.01. Hence, the relationships of the second matrix are quite reliable. However, the first one also gives us additional information for tentative prediction: in it, when W' is high, there are medium and high catch levels with a probability of 80%. In the second matrix, with a probability of 100%, a high level of δ corresponds to low catches.

The mechanism of the links of the initial factors (W' and δ) with the atmospheric transports is revealed by their correlation comparison (Table 12). The table shows that the Earth's rotation rate (δ) is in an inverse relationship with the meridional transports in the first half of the year and during the year, and the modulus of the solar activity anomaly (W') has an even closer relationship (significance level 0.01 in three cases out of five) with the indicated atmospheric transports (realization at the joint influence - δW').

*Table 12.* **Matrix of correlation relations of geo- and helio-physical factors with atmospheric transports within the analyzed SEE water area (series for the period 1972-2001).**

| Transfer factor | δ | δW' |
|---|---|---|
| $A_{10-1}$ | -0,381(< 0,05) | -0,504(< 0,01) |
| $A_{10-2}$ | | -0,364(< 0,05) |
| $A_{10\text{-год}}$ | -0,371(< 0,05) | -0,455(< 0,05) |
| $A_{20-1}$ | -0,398(< 0,05) | -0,477(< 0,01) |
| $A_{20\text{-год}}$ | -0,434(< 0,05) | -0,492(< 0,01) |

A simple conclusion can be drawn from this system of relations: the Earth rotation rate and the modulus of the solar activity anomaly are directly related to the meridional transport from south to north, i.e., to the transport of water from the Antarctic Circumpolar Current (ACC) area to the Peruvian upwelling system. This causes an increase in the inflow of productive waters from high latitudes to the Peruvian upwelling zone, simultaneously with its dynamic amplification, and intensification of macroscale cycles. Thus, the yield of mackerel increases and the conditions of its accumulation in the fishing area improve. Thus, the multiyear forecast can be based on extrapolated factors W' (Khramova et al., 2001) and δ - on the continuation of the multiyear series taking into account the 70-year periodicity.

*Table 13.* **Extrapolated factors and projected catches of jack mackerel in the SEE in absolute (mln t) and relative units**

| Year | W' | δ | St (mln tons) | Assessment |
|---|---|---|---|---|
| 2009 | 16 | 0,94 | 1,92 | H |

| 2010 | 27 | 0,91 | 2,15 | H |
|---|---|---|---|---|
| 2011 | 56 | 0,88 | 2,68 | C |
| 2012 | 52 | 0,86 | 2,64 | C |
| 2013 | 43 | 0,83 | 2,56 | H |
| 2014 | 3 | 0,80 | 1,97 | H |
| 2015 | 27 | 0,77 | 2,40 | H |
| 2016 | 52 | 0,74 | 2,86 | C |
| 2017 | 59 | 0,71 | 3,04 | C |

The extrapolation performed and the indicators in Tables 9 and 11 give us a tentative outlook for the mackerel fishery in the SEEO. The average and low levels of the Earth rotation velocity, taking into account its descending branch after the maximum in 2007, will be characteristic after 2014, therefore, we can assume a steady increase in catches from 2015 to 2016. Some intermediate increase, according to extrapolation of solar activity values and dependencies presented in Tables 10 and 11, can be expected in 2011 and 2012. Approximately the same forecast is obtained by calculations using equation (7). In Table 13 we present the projected catch values assuming the level of exploitation of the raw material base typical for the analyzed period and without taking into account significant climatic changes. In the presence of these factors, the forecast corresponds to the relative characteristics given in Table 10 (H, C, B).

## 7. Arabian Sea*

Changes in the state of the ecosystem in the water area of the western Arabian Sea depending on natural factors presented in the work of S.A. Piontkovski (Piontkovski et al., 2016) include such effects as a decrease in primary productivity, expressed by the concentration of chlorophyll *a*, a decrease in sardine catches, and an increase in the frequency of over-water events off the Omani coast. But when taking into account the general factors of influence, there is a possibility of multi-year forecasting of such changes due to extrapolation of the latter (Bryantsev, 2013). These include: indicators of solar activity (W - Wolf numbers), the module of differences from the average (anomaly - W'), data on changes in the Earth's rotation rate, reflecting climatic changes on the planet (δ) (Sidorenkov and Svirenko, 1989), as well as their products (δW and (5W').

We carried out a correlation analysis of the above factors and the values of Chl. *a* concentration as the main indicator of the ecosystem state, as well as the values of zonal and meridional components of the wind field ($Z_W$ $M_W$)

taken from the above-mentioned work of S.A. Piontkovsky. Its results are summarized in Table 14 and Fig. 5, which show that the abundance of primary production in the western Arabian Sea during the summer monsoon decreases as the zonal component of the wind field increases. The latter, as is known (Kanaev et al., 1975), causes a general anticyclonic circulation of waters in the Arabian Sea and a decrease in the productivity of the water area (except for coastal areas).

The section is co-authored with Y.V. Bryantseva.

*Table 14.* **Correlation coefficients and significance levels of relationships of chl. *a* values averaged over the water area of the western Arabian Sea***

| Argument Function | Zw | W | Δ | δW | δW· |
|---|---|---|---|---|---|
| hl. *a* | -0,466 < 0,01 | | | | -0,353 < 0,01 |
| $M_w$ | | | -0,463 < 0,01 | | -0,411 < 0,01 |
| Zw | | 0,508 < 0,01 | | 0,761 < 0,01 | 0,543 < 0,01 |

* Period of the summer monsoon, meridional and zonal components of the wind field (1960-2013), as well as external factors: solar activity, the Earth Climate Change Indicator and their products (see text for designations).

Thus, against the general background of decreasing primary production in some years, the above effect increases with the intensification of the southwest monsoon.

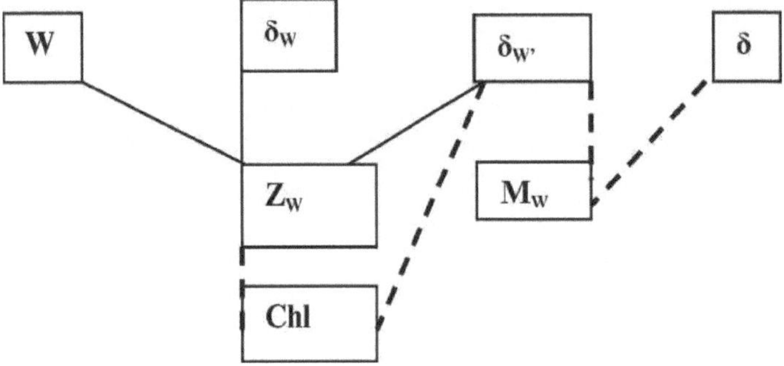


Fig. 5. Schematic diagram of relationships between characteristics and factors of the ecosystem of the western Arabian Sea: solid line - direct relationships; dashed line - inverse relationships (see text and Table 14 for designations) 44

With the help of extrapolation of the listed initial factors, including cyclic fluctuations of 11 (W), 5-6 (W') and 70 (δ) years, respectively, it is possible to forecast the abundance of primary production in the investigated water area and all the consequences presented in the above-mentioned work of C.A. Piontkovsky.

## 8. Azov-Black Sea basin

### 8.1. Runoff of the Dnieper and Danube

Heliophysical indices of solar activity - Wolf numbers and their anomalies $|Wr = W_i - Wcp|$ are correlated with the values of annual runoffs ($km^3$ ) of the Dnieper and Danube rivers in the period 1961-1983 in order to identify their statistically significant relationships for obtaining a multi-year approximate forecast of their runoff. Both series of runoff values were correlated with each other (r = 0.619, significance level 0.01), which suggests that there is a sign of a system of links between hydrometeorological processes in the temperate zone of Europe.

Local climate characteristics were assessed by L.A. Kovalchuk (2012) on the basis of monthly average air temperature and monthly precipitation data in Kiev, registered by the Hydrometeorological Service of Ukraine in the period from 1900 to 2010. With the author's consent, we used these data for the analyzed period 1961-1983, for which we calculated the series of monthly precipitation sums (EOS) and dispersions in the series of daily mean air temperature values for each year (σ(t)).

The results of correlation analysis of the Dnieper and Danube River runoffs with the above series and solar activity indices are presented in Table 15 and Fig. 6. 6. They show the average dependence of the Dnieper and Danube runoffs on solar activity. The mechanism of solar activity energy transfer to the troposphere has not yet been precisely determined. However, the works of a number of researchers present hypotheses of such a connection. The monograph by M.I. Budyko, for example, sets forth a hypothesis of such an influence system (Budyko, 1974).

*Table 15.* **Correlation indices of Dnieper (Dn) and Danube (Du) runoffs with meteorological and heliophysical characteristics (correlation coefficients and significance levels)**

| Argument Function | W | Wr | ΣOc | σ(t) |
|---|---|---|---|---|
| Dn | 0,572 < 0,01 | | 0,524 < 0,05 | |
| Duh | | 0,509 < 0,05 | 0,467 < 0,05 | |
| Xhos | | | | 0,476 < 0,05 |

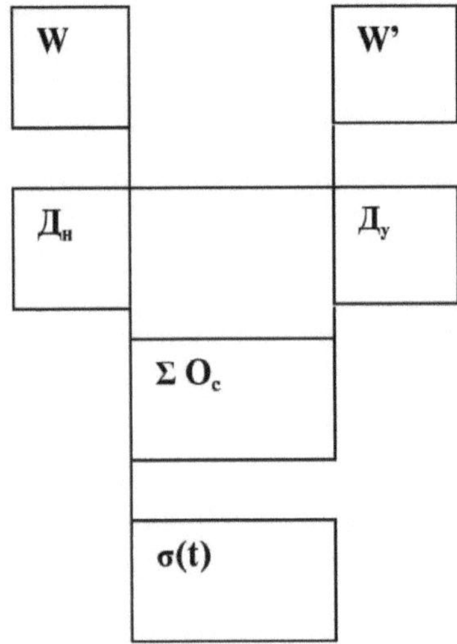


Fig. 6. Scheme of correlation relations of characteristics (see Table 15)

The detected direct connection of the mentioned rivers' runoff with the amount of precipitation and intensity of air temperature fluctuations indirectly indicate an increase in the frequency of cyclone passage in the studied region.

Thus, the regression equation of the relationship between the Dnieper's runoff and solar activity is

Dn = 34 + 0.16W,

with 83% probability (for the Danube - only 74%) can be used for indicative forecast with annual advance. The predictor (W) can be extrapolated, having in mind the periodicity of its fluctuations close to 11 years, or quantitatively - according to the methodology given in Khramov et al. (2001).

**8.2. Black Sea (deep water)**

In recent years, oceanological studies by YugNIRO have established, in particular, that long-term changes in the ecosystem elements of the eastern deep-water part are correlated with atmospheric circulation indicators, thermal characteristics of waters and, all together, with geo- and heliophysical factors external to the sea (Bryantseva et al., 1996; Bryantsev and Bryantseva, 1999). The listed relationships are presented in Table 16. In it, arguments are labeled in column titles, functions - in row titles. The repetition of designations is due to the fact that a certain function is an argument in another relation.

The main object of analysis is the average annual specific biomass of phytoplankton in mg/m$^3$ in the 0-100 m layer according to the data of seasonal surveys of YugNIRO in the eastern (deep -

*Table 16*

**Correlation relationships of deep Black Sea ecosystem elements (correlation coefficients and significance levels)**

| Ecosystem elements (functions) | Arguments | | | | | | | | | |
|---|---|---|---|---|---|---|---|---|---|---|
| | AQO | ф | Tho | TБ | q | E | W | δ | W' | δW |
| ф 1960-1986 | 0,430 <0,03 | - | - | - | 0,519 <0,01 | 0,563 <0,01 | - | - | - | - |
| Bx 1967-1987 | | 0,486 <0,05 | - | -0,488 0,03 | - | - | - | - | - | - |
| B 1985-2000 | - | - | 0,640 0,01 | - | - | - | - | 0,669 0,01 | - | - |
| Aoo 1960-2008 | - | - | - | - | 0,394 0,04 | - | - | - | - | 0,304 0,01 |
| AQI 1960-2008 | - | - | - | - | - | - | - | - | - | 0,538 0,01 |
| Ayu 1960-2008 | -0,404 0,02 | - | - | - | - | - | - | - | - | - |
| To 1985-2000. | - | - | - | - | - | - | - | 0,642 0,01 | -0,243 <0,05 | 0,614* 0,05 |
| TБ 1960 1986 | -0,370 <0,05 | - | 0,631 <0,01 | - | | | - | - | - | - |
| H 1960-1986 | - | - | - | - | - | -0,498 0,04 | - | - | - | - |
| q 1960-1986 | 0,304 0,01 | - | - | - | - | - | - | - | - | - |

Notes. The functions To TБ q - are simultaneously arguments. * For the series 1928-1992.

The final biotic elements of the ecosystem are short-cycle fishes: hamsa (Uh, catches of thousands of tons per trip from 1960 to 1986 (F), published in the Reference Manual (Simonov et al., 1992). The ultimate biotic elements of the ecosystem are short-cycle fish: hamsa (Ukh, catch in thousand tons per trip from 1967 to 1987 and sprat (Shlyakhov et al., 1990); B - catch data from 1985 to 2000 (Shlyakhov and Chashchin, 2000).

The atmospheric circulation indices were obtained by decomposition of daily values of the standard baric field over the water area of the Azov-Black Sea basin into a series by Chebyshev polynomials. The coefficients used are: $A_{00}$ - mean field atmospheric pressure, $A_{01}$ - zonal transport index, $A_{10}$ - meridional transport index. $T_0$ and $T_Б$ - designations of mean annual temperature series, respectively, in the ports of Odessa and Batumi, q - value of irretrievable anthropogenic water consumption in the Azov-Black Sea basin (difference between natural and actual freshwater runoff in cu. km. km according to known data (Nikolenko, Reshetnikov, 1991), H - values of statistical entropy, which is an indicator of diversity measure in phytoplankton community of the eastern (deep-water) part of the Black Sea, E - total impact of natural and anthropogenic factors, obtained by adding indices $A_{00}$ and q after normalizing both series by amplitude (Bryantsev, 2010).

On the one hand, the abundance of phytoplankton in the studied water area directly depends on the peculiarities of atmospheric circulation. Such relationships together with anthropogenic factors (δ, E, H) were described by us earlier (Bryantseva et al., 1996). On the other hand, high atmospheric pressure ($A_{00}$ ) is associated with the predominance of northern atmospheric transport ($A_{10}$ ), which contributes to the strengthening of the cyclonic water circulation and the rise of deep productive waters in the western and eastern Black Sea circulation. The decrease in water temperature ($T_0$ and $T_Б$ ) reflects the intensification of coastal upwelling. Both processes contribute to an increase in the forage base, as well as the yields of hamsa and Black Sea sprat.

Thus, it is acceptable to assume the influence of the solar activity factor and its extreme values (maximum and minimum) on the atmospheric circulation, which are indicated in the monograph by M.I. Budyko (1984). Indeed, when excluding the anomalous values of three years from the phytoplankton series, we obtain significant correlation coefficients of specific biomass with both characteristics - W and W', respectively, 0.534 and 0.438, where $p < 0.05$. The absence of such an effect in the preceding period can be explained by the fact that this time is the 20th solar activity cycle, which has a maximum that is significantly smaller compared to the subsequent one.

The year 1972 marks a minimum in the cycle of changes in the Earth's rotational velocity, reflecting the 70-year period of auto-oscillating climatic fluctuations. The superposition of both factors and anthropogenic influence set a precedent for such a change in the phytoplankton community and the state of the Black Sea ecosystem, which can be labeled as stressful (Bryantseva et al., 1996).

## 8.3. Black Sea ecosystem

In the process of our studies, we revealed successive changes in some biological elements of the Black Sea ecosystem in the period from 1960 to 2000. At the same time, their values in the first half of this period (1960-1972) differed significantly from the second half, starting from 1973 (Mashtakova and Samyshev, 1986; Bryantseva et al., 1996; Bryantsev, 2004; Bryantsev and Kriskiewicz, 2008; Bryantsev and Bryantseva, 2010). Correlations of interannual fluctuations of hydrometeorological and biotic characteristics with solar activity (Wolf numbers) and with the Earth rotation rate (conditional values from 0 to 1) reflecting climatic changes of the planet have been revealed (Sidorenkov, 1989; Sidorenkov, 2004). In the latter case, the fluctuations of some of the analyzed series were correlated with the 70-year periodicity of this geophysical characteristic established by N.S. Sidorenkov and P.I. Svirenko, where the maximum occurred in the mid-30s and the minimum in the mid-70s of the last century (Sidorenkov and Svirenko, 1989).

Against the background of these changes, which can be traced on the ascending segment of the curve of climatic fluctuations (from 1973 to 2006), it is difficult to identify the effects of the Earth's climate warming, especially since its very manifestation in the atmosphere and hydrosphere is complex and ambiguous. According to the literature (Turner, 2009), along with the evidence of the impact of climate warming under the influence of greenhouse gas, a number of factors of influence on the atmosphere and hydrosphere in both circumpolar regions are listed. These include features of orography, land-sea distribution, changes in sea surface albedo, volcanic aerosols and the size of the ozone hole over Antarctica. Interactions of its elements and feedbacks are added to the system of influences.

Naturally, in the case of the Black Sea, the hydrometeorological effects of warming are also distorted 52

due to the "local character" of the region and its remoteness from the Atlantic Ocean. Nevertheless, even here there are signs of the influence of climate warming. They are manifested, in particular, in the long-term series of observations of the surface water layer temperature, where the average temperature and dispersion from 1973 to 2006 significantly differ from the indicators of a number of previous years.

The analyzed series includes water temperature indices in the ports of Odessa (To, 1915-2006) and Batumi ($T_Б$ , 1925-1992), which we used as characteristic for reflecting oceanographic features of the water areas of the northwestern shelf and eastern part of the Black Sea (Bryantsev, 1977).

The atmospheric circulation indices (mean atmospheric pressure ($A_{00}$ ), atmospheric transports) were determined for the period 1960-2006 by decomposing the surface standard baric field within the sea area into a series by Chebyshev polynomials. When analyzing and comparing the series, the first five coefficients reflecting: mean field pressure ($A_{00}$ ) and transfers: from south to north ($A_{10}$ ), from west to east ($A_{01}$ ), from south to north in the western half of the field and from north to south in the eastern half ($A_{20}$ ), from west to east in the southern half of the field and from east to west in the northern half ($A_{02}$ ) were used. Negative values of these coefficients reflect opposite directions.

Hydrobiological data included: average values of the weighted average in the layer (0-100 or 0 - bottom horizon, depending on the depth of the site) of phytoplankton and zooplankton biomass in the area and for the year

(respectively, Fv and /v - in the eastern Black Sea, Fsz and /sz - in the waters of its north-western shelf), obtained during the seasonal standard surveys of the South-NIRO during the periods from 1957-1960 to 1984-1989 (Simonov et al., 1992). The mean biomass values for the main phytoplankton taxa by areas - diatoms (Dsz, Rsz) and peridinium (Dv, Rv) and their ratio (D/Rsz and D/Rv) were also calculated (Bryantseva et al., 1996).

Comparison of series of hydrometeorological and hydrobiological characteristics was performed using the correlation method. Differences of mean and variance in the series were evaluated by the method of null hypotheses (Van der Varden, 1960; Aksyutina, 1968).

Comparisons of all characteristics of the two periods: from 1973 to 1989-2006 and the period preceding it (starting in different years from 1915 to 1960 and ending in 1972) showed differences in the series of periods in terms of mean and variance values. Table 17 presents variants of comparisons where the null hypotheses were rejected with a significance level of at least 0.95 by Student's (t) and/or Fisher's (F) criterion (Aksyutina, 1968; Brooks and Caruthers, 1977).

The longest series of surface water temperature observations in the port of Odessa (1915-2006) reflects the peculiarities of the direct influence of the atmosphere on the state of the waters of the northwestern Black Sea shelf, while the second, "Batumi" series correlated with the average value of the surface water temperature or the thermal background of the entire sea (Bryantsev, 2010).

The data of Table 17 show that the annual average temperature values for the period 1973-2006 did not differ from those for the whole sample and for other time intervals, but the dispersion of this period significantly exceeded this value in the others. This superiority over the values of the period 1915-1934, as a similar ascending segment of the curve of 70-year periodicity of climatic changes, confirms the assumptions that the studied time interval includes the influence not only of the usual indicated autocoletive climatic changes, but also signs of warming of the Earth's climate. The increase of interannual variability in this period is once again confirmed by the presence of extremes in it in the general 90-year series - absolute minimum in 1985 (8.8 °C) and absolute maximum in 1999 (12.9 °C).

The difference of the analyzed period in the Batumi series was expressed in a significant decrease in the thermal background of the Black Sea (Table 17), i.e. it showed its decrease in the epoch of climate warming. The absolute minimum of the 67-year observation series (1987, 15.4 °C) was also recorded during this period. The paradoxical nature of this fact is explained by the fact that the warming effect was manifested in this region in a change in the shape of the atmospheric circulation, as described in the literature (Turner and Overland, 2009). We find evidence for this in the results comparing mean pressure ($A_Q$ J and atmospheric circulation type ($A_{02}$ ). Both of them show a decrease in the frequency of occurrence of the cyclonic-type atmospheric transports.

The weakening and decrease in the frequency of occurrence of the system of westerly transports in the southern part of the sea and easterly transports in the northern part, as well as the predominance of backward transports in the negative phase of the system indicate the strengthening of coastal upwelling in the southern and eastern parts of the sea and the strengthening of cyclonic circulation in the two main Black Sea circles, hence, the inflow of deep productive water to the photic layer. At the same time, the increased frequency of water circulation under westerly transports on the northwestern shelf contributes to the accumulation of discharged river water here, 5

**Significant differences of mean and variance in the studied period**

*Table 17*

| Period series | Characteristics | Rows (periods) | N | X | Σ | Number of series comparison | T | F | Significance level t | Significance level F |
|---|---|---|---|---|---|---|---|---|---|---|

| | | | | | | | | | | |
|---|---|---|---|---|---|---|---|---|---|---|
| 1 | ‰ | 1915-1934 | 20 | H $O_5$ | 0,71 | 1-2 | - | 2,31 | - | >0,95 |
| 2 | | 1973-2006 | 34 | 11,0 | 1,08 | 2-3 | - | 2,40 | - | >0,95 |
| 3 | | 1915-1972 | 58 | H $I_5$ | 0,70 | - | - | - | - | - |
| 4 | | 1915-2006 | 92 | H $O_5$ | 0,85 | - | - | - | - | - |
| 5 | | 1925-1972 | 47 | 16,8 | 0,61 | 5-6 | 3,75 | - | > 0,999 | |
| 6 | $T_6$ | 1973-1992 | 20 | 16,2 | 0,56 | - | - | - | - | - |
| 7 | $A_{00}$ | 1960-1972 | 13 | 15,6 | 0,74 | 7-8 | 3,20 | - | >0,99 | - |
| 8 | | 1973-2004 | 32 | 16,4 | 0,72 | - | - | - | - | - |
| 9 | AQ2 | 1960-1972 | 13 | 0,10 | 0,071 | 9-10 | 4,76 | - | > 0,999 | - |
| 10 | | 1973-2006 | 34 | -0,062 | 0,116 | - | - | - | - | - |
| 11 | | 1960-1972 | 13 | 603 | 212 | I 1-12 | 2,64 | 7,29 | >0,95 | >0,99 |
| 12 | $F_{n3}$ | 1973-1988 | 12 | 1108 | 571 | - | - | - | - | - |
| 13 | | 1960-1972 | 13 | 142 | 60 | 13-14 | 5,13 | - | > 0,999 | - |
| 14 | Hzz | 1973-1988 | 16 | 60 | 46 | - | - | - | - | - |
| 15 | Fc | 1960-1972 | 13 | 82 | 39 | 15-16 | 4,00 | 112,68 | > 0,999 | > 0,999 |
| 16 | | 1973-1984 | 12 | 562 | 414 | - | - | - | - | - |
| 17 | D∏3 | 1957-1972 | 16 | 498 | 185 | 17-18 | - | 8,49 | - | >0,99 |
| 18 | | 1973-1989 | 16 | 810 | 539 | - | - | - | - | - |
| 19 | P∏3 | 1957-1972 | 16 | 76 | 49 | 19-20 | - | 4,77 | | >0,99 |
| 20 | | 1973-1989 | 16 | 134 | 107 | - | - | - | - | - |
| 21 | ∑D $P_{сп3}$ | 1957-1972 | 16 | 573 | 202 | 21-22 | 2,93 | 7,33 | >0,99 | >0,99 |
| 22 | | 1973-1989 | 16 | 1014 | 547 | - | - | - | - | - |
| 23 | $D_c$/P∏3 | 1957-1972 | 16 | 9,40 | 5,78 | - | - | - | - | - |
| 24 | | 1973-1989 | 16 | 5,58 | 5,03 | - | - | - | - | - |
| 25 | ¾ | 1964-1972 | 9 | 66 | 28 | 25-26 | 45,89 | 178,42 | >0,999 | >0,999 |
| 26 | | 1973-1989 | 16 | 442 | 373 | - | - | - | - | - |

| 27 | Pc | 1964-1972 | 9 | 21 | 10 | 27-28 | - | 5,86 | - | >0,99 |
|---|---|---|---|---|---|---|---|---|---|---|
| 28 | | 1973-1989 | 16 | 42 | 23 | - | - | - | — | - |
| 29 | $\sum$D $P_{cc}$ | 1960-1972 | 13 | 74 | 34 | 29-30 | 3,66 | 133 | >0,99 | >0,999 |
| 30 | | 1973-1989 | 16 | 484 | 387 | - | - | - | - | - |
| 31 | $D_c/P_c$ | 1964-1972 | 9 | 3,9 | 2,6 | 31-32 | 2,42 | 10,09 | >0,95 | >0,999 |
| 32 | | 1973-1989 | 16 | ID. | 8,3 | | | | | |

which determines additional inflow of biogenic salts due to outwelling. A significant temperature decrease in the analyzed period of the Batumi series and anomalous increase of phytoplankton biomass in both areas are related to this.

In the eastern part of the sea, the interaction of the listed characteristics is shown in Fig. Fig. 7.

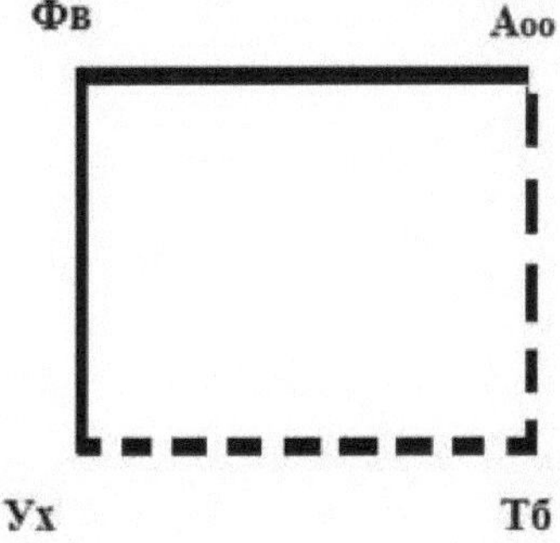


Figure 7. Scheme of Black Sea ecosystem connectivity (Uh - yield of Black Sea anchovy (Shlyakhov et al., 1990)

The average biomass of zooplankton did not change significantly (t and F criteria - not significant). Thus, the state of the ecosystem in this part of the sea can be considered favorable, taking into account the increased yield of Black Sea anchovy, which has the largest share in the catch of all fish species.

In the area of the northwestern shelf, the changes were significant for phytoplankton by both criteria and for zooplankton by mean biomass. The values of phytoplankton biomass more than doubled and zooplankton biomass decreased by the same amount. We assume that in the more eutrophic region of the sea such a situation has developed as a result of the increase in the share of microalgae species that are not fodder for zooplankton, during the outbreak of their development.

*Table 18.* **Correlation coefficients** (R) **and significance levels of relationships of some elements of the Black Sea ecosystem! presented below** (see text for designations)

| Correlation coefficient | Row | | | |
|---|---|---|---|---|
| | Fv-Ao | Fv-Uh | Aooh T 6 | Uh-Tb. |
| R | 0,416<br>< 0,05 | 0,486<br>0,04 | -0,370<br>< 0,05 | -0,488<br>0,03 |

Additional analysis of the dynamics of diatoms and peridinium algae and their biomass ratios showed that in the eastern Black Sea, in addition to a general increase in both taxa, there was an increase in the proportion of diatoms during the period analyzed, resulting in a threefold increase in the Dv/Rv ratio. The changes were significant for both criteria. Large values of Fisher's criterion should be considered as conditional (statistical distribution exceeding normal limits), but a significant increase in mean and variance remains evident. It is noteworthy that the biomass of diatom algae increased almost 7 times, while that of peridinium algae increased only 2 times.

In the area of the northwestern shelf this ratio changed insignificantly, it even decreased, because here the biomass of diatoms and peridiniums increased almost equally. The change in the total biomass of taxa changed significantly by both criteria, but separately by taxa - only by the Fisher criterion. It can be stated that the ecosystem of the north-western shelf has become more productive but less stable compared to the eastern Black Sea.

Thus, the Black Sea ecosystem in its abiotic part changed markedly in 1973-2006. - The interannual variability of water temperature increased and its value decreased in the open part of the sea under the influence of the changed atmospheric circulation system, in particular the reduced recurrence of the cyclonic form. Changes in the biotic part were expressed in a sharp increase in phytoplankton biomass (Vinogradova et al., 1986). As a consequence, the quality of the ecosystem of the region deteriorated on the northwestern shelf, as the anomalous phytoplankton outbreak led to a decrease in the biomass of the second trophic level. The yield of short-cycle fishes, particularly anchovy, increased. Its temporary decrease in 1989-1991 was due to biotic reasons - due to the outbreak of the alien comb mnemiopsis (Ras, 1992).

When comparing the series, first of all, of water temperature indicators in the port of Odessa, which have been kept since 1915, it can be assumed that the perceptible changes in the Black Sea ecosystem occurred when the warming of the Earth's climate was superimposed on the period of the ascending segment of the curve in the auto-oscillatory climate changes with a 70-year period.

### 8.4. Ecosystem of the Sea of Azov

The productivity of the Sea of Azov can be conventionally represented by the total annual catch (harvest, as defined by V.F. Roys (1975). This total biomass is given in FAO statistics and includes 43 species of fish, all mollusks and crustaceans. Between 1970 and 2002, it varied from 7 to 190 thousand tons (Fig. 8, *a* ). In 1970-1987 the annual catch of bioproducts was 127 thousand tons, from 1988 to 2002 only 22 thousand tons. Thus, the productivity of the Sea of Azov, the highest in the World Ocean (Moiseev, 1977), has decreased 6 times. The share of two pelagic fish species - tulka (*Clupeonella cultiventris*) and Azov hamsa (*Engraulis encrasicolus*) was 80-90% before 1988, and after the catastrophic decline (Fig. 8, *c, d*) in some years decreased to 32% (Bryantsev, 20086; Volovik et al., 1996).

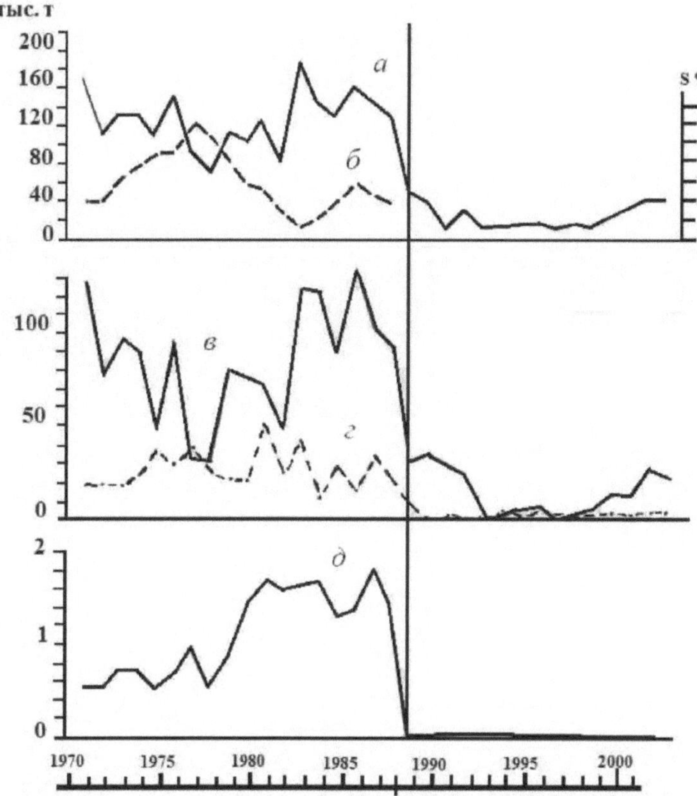

Figure 8. Dynamics of biomass removal from the Sea of Azov (*a*); annual catches of tulka (*c*), Azov hamsa (*d*) and flounder (*e*). Changes in mean salinity ( *b)*

This circumstance divides the period under study into two parts: before 1988, when we have the right to consider as usual the state of the Azov ecosystem depending on natural factors and anthropogenic outflow of river runoff, and after, when regular summer invasions from the Black Sea of an invader - the crested fish Mnemiopsis, with its eating up of the food base of the mentioned fish, caused a decrease in their stock by more than 100 times.

The total catch of hamsa and tulka after 1988 did not rise above the minimum value of the previous period. And since this situation was observed during the next 18 years, the trophic structure of the Azov ecosystem has fundamentally changed. The decrease in the stock of not only the above-mentioned finfish species, but also flounder (Fig. 8, *e* ), indicates that there was a general decrease in the sea yield. Thus, the analysis of natural impacts on the Azov Sea ecosystem and long-term forecasting can be done until 1988.

Our methodology of searching for relationships between biotic indicators and factors external to them is based on correlation comparison of the series of these characteristics at our disposal. The results of the analysis are summarized in Table 19 and Fig. 9.

After 1980, the salinity series was supplemented with the data from observations of South-NIRO. The values of total freshwater runoff in the Azov-Black Sea basin (Q, $km^3$ ) are taken from the literature (Nikolenko and Reshetnikov, 1991). Our calculation of the value of $A_{00}$ , an indicator of the peculiarities of atmospheric transports, is the coefficient of decomposition of the standard baric field into a series by Chebyshev polynomials (mean atmospheric pressure over the basin water area in millibars minus 1000 mb) (Bryantsev, 1990).

*Table 19.* **Relationships between factors of external impacts and parameters of the Azov Sea ecosystem**

| Row | Function | Argument | Correlation coefficient | Significance level | Regression equation |
|---|---|---|---|---|---|
| 1 | $A_{00}$ | δW | 0,364 | < 0,05 | $A_{00} = 0.01\ \delta W+15.8$ |
| 2 | Q | W | 0,467 | 0,01 | $Q = 0.75\ W+138$ |
| 3 | | $A_{00}$ | -0,387 | 0,05 | $Q = 765-36.4\ A_{00}$ |

| | | | | | |
|---|---|---|---|---|---|
| 4 | Qa | δW | 0,552 | <0,05 | Qa = 26.1+δW' |
| 5 | S | δW | -0,491 | <0,01 | S = 12.5-0.014 δW |
| 6 | | δW' | -0,415 | <0,01 | S = 12.26-0.024 δW' |
| 7 | | St | -0,575 | <0,05 | S = 13.6-0.04 St |
| 8 | P | S | -0,524 | 0,02 | P = 208.6-10.836 S |
| 9 | Z | S | -0,574 | 0,02 | Z = 66.86-4.8S |
| 10 | GP | S | -0,540 | 0,02 | CP = 98.46-5.875 S |
| 11 | Wx | W' | 0,499 | <0,05 | $Y_x$ = 17+0,21 W' |
| 12 | W | S | -0,609 | <0,01 | $Y_т$ = 392-24.56 S |
| 13 | Wk | S | -0,700 | <0,01 | $Y_к$ = 6.2-0.4 S |
| 14 | B | S | -0,539 | <0,05 | B = 400.7-21.9 S |

The primary independent factors (arguments) are: the solar activity index (Wolf number - W) and the modulus of its deviation from the mean (W'), as well as changes in the Earth's rotation rate (δ) in relative units obtained over a 70-year period with a maximum in 2005- 2010 (Borovskaya et al., 2005).

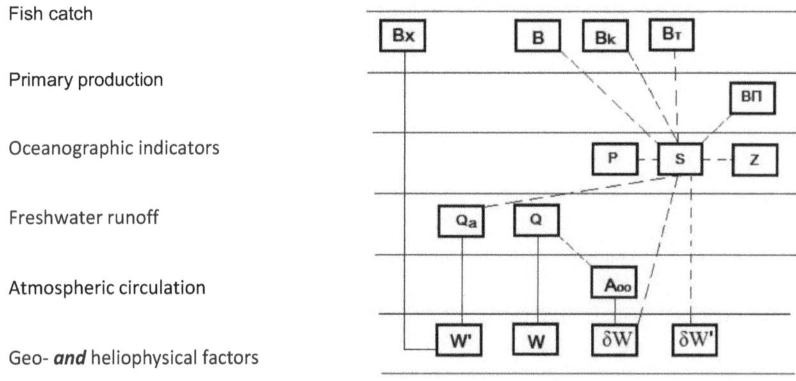


Fig. 9. Functioning of the Azov Sea ecosystem. Schematic diagram of relationships in sequential dependence on geo- and heliophysical factors to the level of total biomass and catches of hamsa, tulka and flounder (1970-1987) (see text for diagram cell designation).

In this scheme of connections the mentioned geo- and heliophysical factors, as well as complex indicators formed by them (δW and δW') are initial in the scheme constructed further in such a sequence: from the indicator of atmospheric circulation peculiarities to the values of water content of years depending on it directly, to oceanographic characteristics of the Azov Sea, to the level of primary production, to the value of annual yield, to the reserves of Azov hamsa, tulka and flounder.

Rows: annual removal of living organisms (harvest - B), annual catch of Azov hamsa (Uh) and seal (Ut) in thousand tons are taken from FAO statistics; values: gross primary production of the Azov Sea in million tons dry weight (β∏), concentration of total phosphorus in mg/m (P). tons are taken from FAO statistics; values of: gross primary production of the Azov Sea in million tons dry weight (β∏), total phosphorus concentration in mg/$m^3$ (P), summer hypoxia area in thousand km2 (Z), freshwater flow in $km^3$ (St) and mean salinity (S‰) are given in the literature (Bronfman, Khlebnikov 1985).

Despite the omission of intermediate links in the identified relationships, due to the lack of links of continuous series of observations of relevant characteristics, the ecological scheme looks regular and not contradictory.

The key indicator of the state of the Azov Sea ecosystem is salinity, which reflects the levels of freshwater runoff in a given and previous years, on which the haline and density structure of waters, their biogenic base and primary production of the sea depend. The last link of the ecological system - the stock of the main commercial fish and flatfish, as well as the total harvest, are correlated with the initial factor (hamsa) or with the intermediate link of the ecosystem (salinity) - the other indicators. It becomes clear that the decrease in the total biomass harvested in the Sea of Azov in 1976 (judging from the increase in salinity, Fig. 6) coincided with the period of low-water years with intensive outflow of Don and Kuban runoff, which caused a decrease in the sea productivity at all trophic levels (Bryantsev, 2008a).

Thus, if the state of the Azov ecosystem stabilizes, long-term forecast will be possible using the specified initial and intermediate (hydrometeorological) characteristics. Then the regression equations of Table 19 can be used as prognostic equations.

### 8.5. Forecast of formation of commercial accumulations of wintering hamsa in the area of Eastern Crimea

In the waters of the eastern coast of Crimea, in some years in the fall, sufficiently stable accumulations of wintering hamsa are formed, which can ensure its successful fishing (Grishin et al., 2013). A significant correlation was also found between cases of such "stopping" of hamsa in the specified area in relatively warm winters (Bryantsev, 20096). The latter characteristic is reflected by the anomaly of average monthly temperatures of the surface water layer in the port of Odessa. In particular, a series reflecting weather conditions in the Azov-Black Sea basin in December (Grishin et al., 2013).

We believe that this indicator is representative, as it is correlated with the data of a similar series in the port of Batumi (correlation coefficient 0.630 at a significance level of less than 0.01), reflecting the thermal background of the sea (see Section 8.1).

An attempt has been made to determine the possibilities of multi-year forecasting of the formation of the above-mentioned accumulations of hamsa off the coast of Eastern Crimea, i.e., the realization of cases of warm winters. The same indicator of the December thermal background in the form of the Odessa series water temperature anomaly can be used as a predictant, as in the work of A.N. Grishin et al. (2013), only for a longer period - from 1928 to 1992 (Ta).

Geo- and heliophysical characteristics were used as predictors: changes in the Earth's rotation rate (δ), the level of solar activity (Wolf numbers - W), the Wolf number anomaly ($W' = W_i$ - W mean) and the products of these values (δW and δW'). As shown above, the first characteristic has a 70-year cycle with a maximum in the mid-30s.

When their cyclicity is taken into account, it is possible to extrapolate geo- and heliophysical characteristics, which, in the presence of their links with indications of the state of marine ecosystems of fishing areas or directly with catches, can become the basis for the methodology of multi-year forecasts of fishing in this region.

On the other hand, their connection with certain indicators of atmospheric circulation and water temperature will allow to reveal the mechanism of changes in the ecosystem and related behavior of the khamsa.

In our case, the relationship of water temperature anomalies in the port of Odessa with the factor δW' is reflected by the correlation coefficient equal to minus 0.261 at the significance level of 0.05. Regression equation

$$Ta = 0{,}8 - 0{,}024\delta W' \qquad (10)$$

can be used for forecasting the level of thermal background in December (Ta) with estimation of the probability of wintering khamsa fishery off the eastern coast of Crimea for the following years when extrapolating the δW' index.

The physical mechanism of formation of conditions for a "warm" December is determined as follows. When comparing the δW' index series from 1960 to 2008 with the values of the baric field decomposition coefficients over the Black Sea area, correlation coefficients (not more than 0.05) were obtained with $A_{00}$ for January (0.371) and $A_{01}$ for the year (0.387). This relationship indicates the dominance of anticyclonic circulation with an increase in the frequency of northerly winds in the eastern half of the Black Sea area. Consequently, under cyclonic circulation, southerly winds and warm winters will prevail here.

According to the obtained prognostic equation, the most favorable conditions for wintering of hamsa off the eastern coast of Crimea are created at the level of the thermal index (Ta) in December 0.8 (zero values of the δW' index). A part of hamsa stayed for wintering in this area in 2011 and 2012, when the Ta factor, according to our data, passed to a positive value and had a value of 0.7 in 2010. Consequently, the tentative forecast can be presented in a dichotomous form: a positive value of the thermal index indicates the probability of stopping a part of the wintering hamsa migrating to the southeastern part of the Black Sea in the area of the eastern coast of Crimea, while a negative value indicates the absence of such.

**8.6.** The fluctuations of **Black Sea sprat** (Ush) catches (Zuev et al., 2003; Eremeev and Zuev, 2005) have noticeable signs of coincidence with our values of the complex factor δW (Table 20), obtained by multiplying solar activity indices (Wolf number) and changes in the Earth's rotation rate (δ).

Both of these geo- and heliophysical factors were found to be significant in our studies in identifying relationships with catch rates and indices reflecting atmospheric circulation patterns in many fishing grounds and in the Black Sea. In turn, prevailing wind transports are associated with known systems of surface water currents in the areas and with the resulting advection phenomena of certain water masses with the formation of frontal zones, upwelling in coastal zones and topogenic eddies. Hydrometeorological prerequisites of biotic and fishery productivity were determined through such an integrated system.

*Table 20.* **Black Sea sprat catches (thousand tons) and factor δW**

| Years | 2000 | 2001 | 2002 | 2003 | 2004 | 2005 | 2006 | 2007 | 2008 | 2009 | 2010 | 2011 |
|---|---|---|---|---|---|---|---|---|---|---|---|---|
| Ush | 10,2 | 19,5 | 21,4 | 13,4 | 12,3 | 17,8 | 14,6 | 11,4 | 15,3 | 18,7 | 20,2 | 20,8 |
| δW | 96 | 92 | 89 | 57 | 36 | 38 | 15 | 8 | 3 | 83 | 105 | 108 |

It is assumed that the chosen initial factors condition the above assumptions. If fluctuations in the Earth's rotation rate (δ) can be explained by known climatic changes in the atmosphere and water circulation, then annual values of solar activity (W) are used in hydrometeorology and in other studies, in particular by A.L. Chizhevsky (1973), formally, according to the "black box" principle. When determining correlations of such indices with catches, or with other indicators of the state of ecosystems of fishing areas, we can apply these geo- and heliophysical characteristics to forecast the success of fishing for a year or more, since their known cyclicity (70 and 11 years, respectively) determines the possibilities of extrapolation.

The performed correlation of catches with our indices showed the following results: the correlation coefficient with the value of solar activity is equal to 0.490, and with "correction" for general climatic changes (δ) it is increased to 0.532, the significance level of which does not reach the critical level (0.576, accepted 0.05) by the value of 0.04. However, a more "coarse" indicator can be used for multi-year forecasts, for example, when estimating catches in three gradations: low (H), average (C) for the analyzed period and high (H). This division can be accomplished using the methodology given in the literature (Brooks and Caruthers 1977) and in Table 21.

*Table 21.* **Discharges of values of annual catches of Black Sea sprat by the Ukrainian fleet (Ush, thousand tons) and factor indicator δW**

| | Gradations | | |
|---|---|---|---|
| Row | H | C | B |
| Ush | < 15,5 | 15,5-19,6 | > 19,6 |
| δW | < 38 | 38-91 | > 91 |

When we distribute catch values by gradations of the factor δW, we obtain a matrix of correspondences of their levels in the form of relative probabilities (Table 22).

*Table 22.* **Matrix of correspondence between levels of indicators and Ush and δW**

| Row | δW | | |
|---|---|---|---|
| Ush | H | C | B |
| H | 0,80 | 0,33 | 0,25 |
| C | 0,20 | 0,33 | 0,25 |
| B | 0 | 0,33 | 0,50 |

The matrix shows that at low values of the external factor δW the probability of low catches reaches 80%, and at high values the probability of medium and high catches is 75% in total. At medium values of the factor, the probability of all catch levels is the same.

Hydrometeorological prerequisites of fluctuations in sprat yield, essentially determining its stock at commercial age and formation of aggregations during feeding migrations, are as follows. The complex factor δW used by us is significantly correlated with the indices of atmospheric transports. This is shown by the coefficient of decomposition of the baric field over the water area of the Azov-Black Sea basin - $A_{02}$ (correlation coefficient 0.358, significance level less than 0.05). Physically, this means (with the indicated feedback) a weakening of the recurrence of east-west wind transport in the northern half of the Black Sea. In this case, the intensity of river water transport from the water area of the north-western Black Sea shelf decreases, which increases the primary production and feeding base of sprat here, as well as the accumulation effect in the frontal zones.

Thus, after extrapolation of the components of the prognostic factor (δW) values, we can identify a trend in the change of the fishing success of the mentioned fish species. After reaching the maximum in the first decade of the XX century the value of the index, its subsequent decrease after 2011 will cause, in general, a negative trend in the fishing success of the Black Sea sprat, at least from 2014 to the early 20s.

### 8.7. Forecast of Overcast Events on the North-West Black Sea Shelf

The intensity and duration of overwash events on the northwestern Black Sea shelf in the summer and fall, which have increased since the mid-1970s, have drawn special attention to this environmental problem (Zaitsev et al., 1985; Belyaev and Kondofurova, 1990; Bryantsev et al., 1991). Some scientists attributed the increased hypoxia in the bottom horizons to the influence of chemical processes and, consequently, anthropogenic eutrophication, while others attributed it to the transformation of the salinity field, density and the water circulation system during the weaning and seasonal redistribution of the Dnieper's flow, considering the presence of oxidizable organics to be a necessary condition, which is realized in hypoxia and frost only in the presence of a blocking layer with a high vertical density gradient and reduced shear currents.

To date, there is no precise definition of the mechanism of oxygen deficiency in this region. At the same time, there is a need for a long-term (a year or more) forecast of the phenomenon, at least in the most rough estimates, since it has already led to the death of 70% of mussel stocks and a 20-fold decrease in phyllophora biomass.

Therefore, we assume that the cause of stress changes in the ecosystem of the northwestern shelf water area is a superposition of anthropogenic and natural factors that lead to changes in the hydrostructure and eutrophication of water with subsequent destabilization and qualitative changes in its biotic part. To the former we refer irretrievable water consumption expressed by the difference between natural and actual runoff (Nikolenko, Reshetnikov, 1991). In 1983, this difference reached 39 km$^3$ . Such withdrawal of river water, generally insignificant in comparison with inter-annual flow fluctuations, is accompanied by its seasonal redistribution, i.e. delay in reservoirs during the natural flood period and subsequent discharge with changed physical (temperature) and chemical (as a consequence of production processes) properties.

The natural factors include the predominance of southern and southwestern winds in the summer period, which distort the characteristic cyclonic circulation on the northwestern shelf and develop the anticyclonic circulation, causing the inflow of water from the main rivers of the region to the limits of the specified water area (Andrusovich et al., 1994; Moskalenko et al., 1994). The usual mean climatic situation (Bryantsev, 1987) is expressed by the predominance during this period of atmospheric transports (AT) of the northeastern group, the development of a cyclonic vortex determining the increase in the current shift and the southward transport of river water in a relatively narrow strip along the western coast.

The influx of eutrophic waters into the shelf limits causes an increase in organics and a decrease in transparency. At the same time, stratification increases and current shift decreases, since the usual summer change of the water circulation system does not occur. Thus, the presumed preconditions of the overwintering coincide for both versions. Therefore, we can consider indirect empirical relationships of the areas of freezes with external factors and try to find asynchronous relationships for tentative forecasting.

The possibility of analyzing interannual variability in the state of bottom biocenoses and overwash on the northwestern shelf first appeared after Y.P. Zaitsev published the values of areas subjected to these processes in 1973-1990 (Zaitsev, 1992). Since this is the value of the main series 73 prsdictant in this study, we consider it necessary to duplicate it (Table 23).

*Table 23.* **Areas of the north-western Black Sea (thousand km$^2$ ) affected by hypoxia and bottom animals in 1973-1990 (Zaitsev, 1992).**

| Year | S | Year | S | Year | S | Year | S | Year | S |
|---|---|---|---|---|---|---|---|---|---|
| 1973 | 3,5 | 1977 | 11 | 1981 | 17 | 1985 | 5 | 1989 | 20 |
| 1974 | 12 | 1978 | 30 | 1982 | 12 | 1986 | 8 | 1991 | 40 |

| 1975 | 10 | 1979 | 15 | 1983 | 35 | 1987 | 9 | |
|---|---|---|---|---|---|---|---|---|
| 1976 | 3 | 1980 | 30 | 1984 | 10 | 1988 | 12 | |

The mean annual values of solar activity (W - Wolf numbers, W" - the same values with a shift by 1 year) and values of actual freshwater runoff (Q, $cu^3$ ) (Nikolenko and Reshetnikov, 1991), as well as a number of atmospheric circulation indices were used as an index of external preconditions for anomalous hypoxia development. The latter were calculated by us using the following methodology.

Daily values of atmospheric pressure from 1960 to 1993 were taken from surface baric maps at the nodes of a standard grid completely covering the Black Sea area: from 40° to 46°N through 2 degrees and from 28° to 40°E through 4 degrees. By decomposing these baric fields into a series using Chebyshev polynomials, dividing the values of the obtained coefficients reflecting the intensity of zonal and meridional transports into three equally probable ranges (H - low, C - medium and H - high values) and taking into account combinations of the latter, estimates of the AP types (homonymous with 8 rhumbas, type 9 - low-gradient field) were obtained for each investigated 34-year series (Table 24) (Bryantsev, 1996).

*Table 24.* **Atmospheric transports over the Black Sea area**

| Assigned numbers | Carryover (rhumbas) | Combining ranges of zonal and meridional transport coefficients |
|---|---|---|
| 1 | CB | HH |
| 2 | B | NS |
| 3 | SE | NOT |
| 4 | C | SI |
| 5 | - | SS |
| 6 | Ю | CB |
| 7 | NW | Vn |
| 8 | 3 | VS |
| 9 | Yuz | BB |

Denoting each day of the analyzed period by the symbol of the AP type or this number from Table 24, we can calculate the recurrence of the types in days for each month and year, as well as the mean annual values and anomalies:

$$a_{ij} = P_{ij} - N_{ij}, \qquad (11)$$

where $P_l$ j - recurrence in days of type j (j = 1, 2, 3 ... 9), in the l-th month,

$N_{lj}$ - monthly norm.

The index of dominance of the northeastern AP group (NE, B, SE, C rhumbas) over the southwestern group (SW, NW, 3, SW) in specific months is obtained from the difference of the sums of the corresponding anomalies:

$$B_i = \sum_{j=1}^{j=4} a_{ij} - \sum_{j=6}^{j=9} a_{ij}, \quad (12)$$

and the total anomaly can be calculated as the sum of anomalies of all types regardless of sign:

$$A_i = \sum_{j=1}^{j=9} a_{ij} \quad (13)$$

and for 1 year ( $A_k$ ) also in number of days:

$$A_k = \sum_{i=1}^{i=XII} A_i \quad (14)$$

The anthropogenic factor is represented as a series of irrecoverable water consumption values (q) calculated for the period 1960-1986 as the difference between the volume of natural and actual freshwater runoff.

Correlation and multiple correlation were used to identify relationships between the listed series. Correlations with a significance level of $p < 0.05$ were accepted and included in the summary table. The values of the compared series were determined by the number of years of observations, except for the series of solar activity and recurrence of southern transports in July, as well as the index of prevalence of northeastern transports in the same month, because these relationships were significant in the period 1981-1993.

The results of the analysis are presented visually in the form of a general scheme of relations of the listed characteristics (Fig. 10) and their quantitative estimates (Table 25). The correlation coefficients given in it, although selected according to the specified criterion of significance level, are low in most cases, so there was a problem in using the obtained relationships for forecasting.

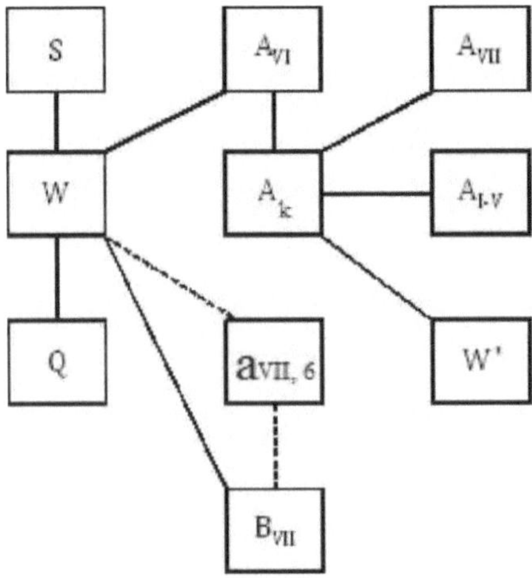


Fig. 10. Scheme of relationships between the level of hypoxia development on the northwestern Black Sea shelf and parameters of external influences.

The dashed line indicates feedbacks

In addition, the relations of the atmospheric circulation indices among themselves have a formal character, since the value of the annual total anomaly (Ak) includes the values of monthly anomalies (Ai), and the criterion of the dominance of northeastern transports in July - the value of the anomaly of the southern AA in the same month. However, we adopted the conditions for assessing the relationships at the level of presence of their signs and forecasts - at the level of determining the trend or years-analogs.

*Table 25.* **Quantitative indicators of the system of links between the level of hypoxia development on the north-western shelf of the Black Sea and parameters of external influences**

| Row | Connections | Coefficient, correl, | Significance level | Number of members in rows | Year of series comparison |
|---|---|---|---|---|---|
| 1 | W-S | 0,624 | 0,006 | 18 | 1973-1990* |
| 2 | W-Q | 0,476 | 0,01 | 27 | 1960-1986 |
| 3 | W-Q $6_{YII.}$ | -0,580 | 0,04 | 13 | 1981-1993 |
| 4 | W-YII | 0,813 | 0,001 | 13 | " -- |
| 5 | W-Ak | 0,391 | 0,02 | 34 | 1960-1993 |
| 6 | W-$A_{YI}$ | 0,394 | 0,02 | 34 | И |
| 7 | W-$A_I$ | 0,400 | 0,02 | 34 | Н |

| | | | | | |
|---|---|---|---|---|---|
| 8 | Ak-$A_{YI}$ | 0,380 | 0,03 | 34 | " -- |
| 9 | Ak-$A_{YII}$ | 0,390 | 0,02 | 34 | " -- |
| 10 | Ak-$A_{I}$ | 0,537 | 0,001 | 34 | H |
| 11 | Ak-$A_{II}$ | 0,380 | 0,03 | 34 | " -- |
| 12 | Ak-$A_{III}$ | 0,524 | 0,002 | 34 | " -- |
| 13 | Ak-$A_{IY}$ | 0,545 | 0,001 | 34 | H |
| 14 | Ak-$A_{Y}$ | 0,437 | 0,01 | 34 | " -- |
| 15 | Ak-W | 0,331 | 0,05 | 34 | " -- |
| 16 | W'-A, | 0,455 | 0,007 | 34 | " -- |
| 17 | Q -□$_{YII\ 6YII}$ | -0,558 | 0,006 | 34 | II |

* Within the period of significant relationships.

The obtained system with all its weak and formal connections forms a consistent and physically conditioned network, which allows us to determine the intermediate links of the relationship between the magnitude of the areas of freezing and solar activity.

As it has been repeatedly noted in the literature, with increasing solar activity, water content of years increases and freezing events are observed more often. Fig. 10 shows that solar activity is related to atmospheric circulation indices of all species:

— with the level of the annual total anomaly and especially with the June level, i.e. with the period of the beginning of formation of conditions for the development of summer hypoxia;

— with predominance of northeastern APs over southwestern ones;

— with the anomaly of the southern transport in July (feedback), which is a prerequisite of unfavorable anticyclonic circulation of water in the water area of the northwestern shelf, determining the increased transport of river water to its limits.

In addition to the relationship of the area of freezing with solar activity and the peculiarities of the hydrometeorological regime created by it, signs of such a relationship with irretrievable water consumption were revealed (Rsq = 0.380), but the usual limitation on the level of significance (here only 0.18) does not allow including this relationship in the general scheme.

Taking into account the results obtained, it can be concluded that the predominance of the south-western group of atmospheric transports, especially southern transports, over the Black Sea water area during the summer period causes an increased probability of ecological crises in the form of extensive freezes on the north-western shelf (Bryantsev and Kochergin, 2008). Under such peculiarities of atmospheric circulation, the waters of the rivers of this region are transported to the limits of this water area and create specific conditions of trophicity, water transparency, as well as their density stratification and shifting currents. The increased, as compared to a multiyear average, frequency of such transfers is observed at high values of solar activity.

Since the Wolf numbers can be extrapolated due to a well-defined 11-year cycle, we obtain the possibility of approximate forecasting of the intensity of overcast events with a year or more in advance. The area of hypoxia can be calculated by the formula:

$$S = 6{,}5 + 0{,}12\ W. \qquad (15)'$$

The assurance of the predictive equation increases when the irrecoverable water use indicator is included (multiple correlation coefficient 0.301):

$$S = 0{,}1\ W + 0{,}5\ q - 5{,}2. \qquad (16).$$

However, this characterization can be determined very roughly, e.g. by the average level in recent years.

## CONCLUSION

On the examples of some fishing areas of the World Ocean, we have shown the dependence of cyclic changes in fishing success on external factors: solar activity and global climatic fluctuations. The latter are also correlated with other indicators of the state of these ecosystems: characteristics of phytoplankton communities, stocks and concentration of Antarctic krill, and occurrence of hypoxia.

We believe that this approach can be used to search for links between solar activity and other hydrometeorological and biological processes on the Earth in order to obtain the possibility of their multi-year forecast, as shown in the works of A.L. Chizhevsky, I.V. Maximov, J.R. Herman, and others.

## LITERATURE

*Aksyutina Z.M.* Elements of mathematical evaluation of observation results in biological and fishery research. - M.: Pishch. promst, 1968. - 288 c.

*Andrusovich A.I., Mikhailova E.N., Shapiro N.V.* Numerical model of water circulation in the north-western part of the Black Sea // Marine Hydrophys. zhurn. - 1994. - № 5. - C. 28-42.

*Belyaev V.I., Kondofurova N.V.* Mathematical modeling of ecological systems of the shelf. - Kiev: Nauk. dumka, 1990. - 240 c.

*Borovskaya R.V., Panov B.H., Spiridonova E.O., Lexikova L.A.* Relation of benthic hypoxia and fish starvation in the coastal part of the Azov Sea // Environmental Control Systems. - Sevastopol: MGI NASU, 2005. - C. 320-328.

*Bronfman A.M., Khlebnikov E.P.* Sea of Azov. - L.: Gidrometeoizdat, 1985. - 272 c.

*Brooks K., Caruthers N.* Application of statistical methods in meteorology. - L.: Gidrometeoizdat, 1977. - 352 c.

*Bryantsev V.A.* About the possibility of seasonal forecast of the Black Sea thermal background // Express-information of the Central Scientific Research Institute of the Black Sea. - M., 1977. - Issue. 8. - C. 1-5.

*Bryantsev V.A.* Methodical recommendations on hydrometeorological forecasting for the main objects of fishing in the Black Sea. - Kerch, 1987. - 42 c.

*Bryantsev V.A.* Atmospheric circulation as a basis for long-term fishery forecasts (on the example of the Black Sea)) // Plenary reports of the 8th All-Union Conf. on Fishery Oceanology. - M., 1990. - C. 173-180.

*Bryantsev V.A.* Information in the form of total anomalies of atmospheric circulation and its impact on the Black Sea ecosystem // Dop. NASU. - 1996. - № 9. - C. 163-168.

*Bryantsev V.A.* External preconditions of perennial changes in the Black Sea ecosystem // Rib. gosp. ukrashi. - 2001. - **17**, № 6. - C. 22-23.

*Bryantsev V.A.* Signs of changes in the state of the ecosystem of the north-western shelf of the Black Sea in connection with the warming of the Earth's climate // Rib. gosp. ukrashi. - 2004. - **34**, № 5. - C. 49-51.

*Bryantsev V.A.* Global processes determining the state of the Azov-Black Sea basin ecosystem // Mat. of IV Intern. IV Intern. conf. (Kerch, October 8-9, 2008). - Kerch: YugNIRO, 2008a. - C. 3-7.

*Bryantsev, V.A.* Stress transformation of the Azov Sea ecosystem and possibilities to restore its fish productivity (in Russian) // Rib. gosp. ukrashi. - 2008б. - **57**, № 4. - C. 3-7.

*Bryantsev V.A.* Factors determining the success of mackerel fishing in the southeastern part of the Pacific Ocean // Proc. of YugNIRO. YugNIRO. - 2009a. - **47**. - C. 206-212.

*BryantsevV .A.* Hydrometeorological conditions of wintering of the Black Sea anchovy off the coast of Crimea // Rib. gosp. ukrashi. - 2009б. - **7**. - C. 8-9.

*Bryantsev V.A.* Climatic changes in the ecosystem of the Azov-Black Sea basin // Modern problems of ecology of the Azov-Black Sea region: Mat. international conf. - Kerch: YugNIRO, 2010. - C. 3-7.

*Bryantsev V.A.* Statistical model of commercial productivity on the example of Black Sea sprat // Mat. VII International Conf. (Kerch, 2012). - Kerch: YugNIRO, 2012. - C. 26-28.

*Bryantsev V.A., Bryantseva Y.V.* Signs of global warming impact on the Black Sea ecosystem // Environmental security of coastal and shelf zones and integrated use of shelf resources.- Sevastopol, 2010. - Vyp. 22. - C. 191 - 197.

*Bryantseva Yu.V., Bryantsev V.A., Kovalchuk L.A., Samyshev E.Z.* To the question of long-term changes in the biomass of diatoms and peridinium algae of the Black Sea in connection with atmospheric transport // Marine Ecol. - 1996. - Vol. 45. - C. 13-18.

*Bryantsev V.A., Bryantseva Y.V.* Multiyear changes in phytoplankton of the deep-water part of the Black Sea in connection with natural and anthropogenic factors // Marine Ecology. - 1999. - Vol. 49. - C. 24-28.

*Bryantsev V.A., Kriskiewicz L.V.* State of the Black Sea ecosystem: empirical assessment, forecasting possibilities // Modern problems of ecology of the Azov-Black Sea region: Mat. III Intern. III Intern. conf. (Kerch, October 10-11, 2007). - Kerch: YugNIRO, 2008. - C. 83-89.

*Bryantsev V.A., Kochergin A.T.* Estimation of probability of occurrence of pre-freeze and freezing situations in the Sea of Azov // Modern problems of ecology of the Azov-Black Sea region: Mat. of III Intern. III Intern. conf. (Kerch, October 10-11, 2007). - Kerch: YugNIRO, 2008. - C. 98-101.

*Bryantsev V.A., Rebik S.T.* Prerequisites of commercial productivity in the area of the Patagonian shelf // Proc. of YugNIRO. YugNIRO. - 2011. - **49**. - C. 199-202.

*Bryantsev V.A., Trotsenko B.G.* Prerequisites of commercial productivity in some areas of the Southern Ocean // Proc. of YugNIRO. YugNIRO. - 2010. - **48**. - C. 119-124.

*Bryantsev V.A., Fashchuk D.Y., Finkelstein M.S.* Signs of trend changes in the

Black Sea hydrostructure // Variability of the Black Sea ecosystem / Edited by M.E. Vinogradov. Vinogradov. - Moscow: Nauka, 1991. - C. 89-93.

*Budyko M.I.* Climate changes. - L.: Gidrometeoizdat, 1974. - 280 c.

*Budyko M.I.* Evolution of the Biosphere. - L.: Gidrometeoizdat, 1984. - 488 c.

*Van der Varden B.L.* Mathematical Statistics. - M.: IL, 1960. - 434 c.

*Vinogradova L.A., Mashtakova G.P., Derezyuk N.V.* Successional changes in phytoplankton of the north-western part of the Black Sea. - M.: IOAN, 1986. - C. 170-176.

*Volovik S.P., Makarov E.V., Semenov A.D. The* Sea of Azov: Is it possible to get out of the ecological crisis // Main problems of fishery and protection of fishery water bodies of the Azov basin. - Rostov-on-Don: Polygraph, 1996. - C. 115-125.

*Herman J.R., Goldberg R.A.* Sun, weather and climate. - L.: Gidrometeoizdat, 1981. - 320 c.

*Grishin A. H., Serbin V. A. A., Kriskiewicz L. V.* Climatic prerequisites for the formation of wintering aggregations of hamsa (*Engraulss engrasicolus* (L.) off the eastern coast of the Crimea // Mat. VIII Intern. conf. - Kerch: YugNIRO, 2013. - C. 4-7.

*Eremeyev V.H., Zuev G.V.* Fish resources of the Black Sea: long-term dynamics and management prospects // Marine Ecol. zhurn. - 2005. - **4**, № 2. - C. 5-21.

*Zaitsev Y.P.* Ecological state of the Black Sea shelf zone off the coast of Ukraine // Hydrobiol. *zhurn*. - 1992. - **28**, № 4. - C. 3-18.

*Zaitsev Y.P., Bryantsev V.A., Fashchuk D.Ya.* Ecosystems of the North-West Black Sea shelf under anthropogenic impact // Anthropogenic eutrophication of natural waters of the Black Sea. - Chernogolovka, 1985. - C. 49-72.

*Zuev G.V., Bondarev V.A., Samotoy Y.V.* Multiyear dynamics of the Black Sea sprat (*Spratus phalerius* Risso) fishery in Ukrainian waters // Rib. gosp. Ukra. - 2003. - № 1. - C. 15-23.

*Kramer G.* Mathematical methods of statistics. - Moscow: Mir, 1975. - 468 c.

*Kondratovich K.V.* Long-term meteorological forecasts in the North Atlantic. - L.: Gidrometeeoizdat, 1977. - 184 c.

*Kudryavaya K. I. I., Seriakov E.I., Skriptunova L.I.* Marine hydrometeorological forecasts. - L.: Gidrometeoizdat, 1974. - 310 c.

*Maksimov V.I.* Geophysical forces and ocean waters. - L.: Gidrometeoizdat, 1970. - 447 c.

*Marty Yu.Yu., Martinsen S.V.* Problems of formation and use of biological productivity of the Atlantic Ocean. - M.: Pishch. Promst, 1969. - 268 c.

*Maslennikov V.V.* Climatic fluctuations and marine ecosystem of the Antarctic. - Moscow: VNIRO Publishing House, 2003. - 295 c.

*Matishov Yu.Yu., Dzhenyuk S.L., Moiseev D.V., Zhichkin A.P.* Climatic changes in the marine ecosystems of the European Arctic // Problems of the Arctic and Antarctic. Arctic and Antarctic. - 2010. - **86**, № 3. - C. 7-21.

*Mashtakova T.P., Samyshev E.Z.* Structure of plankton communities in the Black Sea and its changes in 1960-1983 // Research of the Black Sea pelagial ecosystem. - M.: IOAN, 1986. - C. 238-240.

*Moiseev P.A.* Fisheries production of the World Ocean and its utilization // Ocean Biol. - M.: Nauka, 1977. - VOL. 2. - P. 289-314.

*Moskalenko L.V., Osadchiy A.S., Titov V.B.* Vertical structure and spatial and temporal variability of hydrophysical fields in the centers of quasi-stationary cyclonic cycles of the Black Sea // Oceanology. - 1994. - **34**, № 3. - C. 349-355.

*Nikolenko A.V., Reshetnikov V.I.* Studies of multiyear variability of the balance of fresh water of the Black Sea // Vod. res. - 1991. - № 1. - C. 20-28.

*Pankratova T.M., Bryantsev V.A.* On the six-year periodicity in phytoplankton production in the deep-water part of the Black Sea // Rib. gosop. ukrashi. - 2002. № 7. - C. 34-37.

*Fishery* description of the Southwest Atlantic. - M.: GU navigation and oceanography, 1984. - 145 c.

*Fishery* description of the Southeast Pacific Ocean area. - M.: GUNIO, 1985. - 154 c.

*Ras T.S.* Fish resources of the Black Sea and their changes // Oceanology. - 1992. - **32**, Vol. 2. - C. 293-301.

*Royce V.F.* Introduction to fishery science. - M.: Pishch. promst, 1975. - 272 c.

*Samyshev E.Z.* Antarctic krill and structure of the planktonic community in its area. - M.: Nauka, 1991. - 166 c.

*Sidorenkov N. S.* Speed of Earth rotation. - http://vivoco. rsl. ru IvvI (RAN) 2004/FLUCT. HTM. - 12 c.

*Sidorenkov N.S., Svirenko P.I.* Multiyear changes of atmospheric circulation and climate fluctuations in the first natural synoptic region // Long-period variability of the environment and some issues of fishery forecasting. - M.: VNIRO, 1989. - C. 59-71.

*Simonov V.I., Ryabinin A.I., Gershanovich D.E.* Hydrometeorology and hydrochemistry of the seas of the USSR. Black Sea. T. 4, vol. 2. - St. Petersburg: Gidrometeoizdat, 1992. - 220 c.

*Khramova M.N., Krasotkin S.A., Kononovich E.V.* Forecasting of solar activity by the method of phase averages // Researched in Russia. 1169. http/zhurnal/ apo. ru// articles/ 2001/ 107. pdf.

*Chizhevsky A.L.* Earth Echo of Solar Storms. - Moscow: Mysl, 1973. - 348 c.

*Shlyakhov V.A., Chashchin A.K., Korkosh N.I.* Fishing intensity and dynamics of the Black Sea hamsa stock // Biological Resources of the Black Sea. - M.: VNIRO, 1990. - C. 93-102.

*Shlyakhov V.A., Chashchin A. K.* On the state of stocks of the main commercial fishes of the Azov and Black Seas in 2000 and prospects of their fishing in 2002 // Proc. of YugNIRO. YugNIRO. - 2000. - **45**. - C. 11-20.

*Bibik V.A., Yakovlev V.N.*. The status of Krill (*Euphausia superba* Dana) resources in CCAMLR statistical divisions 58.4.2 and 58.4.3 from 1988 to 1990. Results of acoustic surveys // WG - Krill. - 1990. - **90,** N 17. - P. 163-173.

*Cushing D.H.*. Population production and regulation in the sea: a fisheries perspective. - Cambridge: Print. Great Brit. Univ. Press, 1995. - 348 p.

*FAO* year book. - 2005. - **100,** N 1. - P. 205-231.

*Piontkovski S.A., Bastin Y. Queste.* Decadal changes of the Western Arabian Sea ecosystem |// Int. Aquat Res. - 2016. - **8**. - P. 49- 64. doi: 1007/S 40071-016-0124-3.

*Turner J., Overland I.* Contrasting climate change in the two polar regions // J. Compilation (Blackwell Publ. Ltd. Compilation (Blackwell Publ. Ltd.). - 2009. - P. 146-164.

Printed by Books on Demand GmbH, Norderstedt / Germany